Communications
in Computer and Information Science 440

T0212825

Vladimir Golovko   Akira Imada (Eds.)

Neural Networks and Artificial Intelligence

8th International Conference, ICNNAI 2014
Brest, Belarus, June 3-6, 2014
Proceedings

 Springer

Volume Editors

Vladimir Golovko
Brest State Technical University
Moskowskaja 267
224017 Brest, Belarus
E-mail: gva@bstu.by

Akira Imada
Brest State Technical University
Moskowskaja 267
224017 Brest, Belarus
E-mail: akira-i@brest-state-tech-univ.org

ISSN 1865-0929   e-ISSN 1865-0937
ISBN 978-3-319-08200-4   e-ISBN 978-3-319-08201-1
DOI 10.1007/978-3-319-08201-1
Springer Cham Heidelberg New York Dordrecht London

Library of Congress Control Number: 2014940983

Typesetting: Camera-ready by author, data conversion by Scientific Publishing Services, Chennai, India

Printed on acid-free paper

Springer is part of Springer Science+Business Media (www.springer.com)

Preface

The International Conference on Neural Network and Artificial Intelligence (ICNNAI) started its history in October 1999 in Brest, Belarus. Since then, conferences have been held in 2001, 2003, 2006, 2008, 2010, 2012, with each one evolving to be better than its predecessors. The conference in 2014 was the eighth conference in this series.

In October 2013, with still 8 months to go, while planning how to further develop the conference this time, we received an e-mail from Springer with the suggestion of publishing our proceedings in Springer's CCIS series. It was a pleasant surprise, as this would be a good chance for us to make a big step toward the future. Of course, we applied by sending the required information. It was conditionally accepted after what they wrote was "a tough discussion in the editorial board." The conditions seemed quite demanding but we agreed. Why not? An international conference is not "a meeting of birds of a feather" nor "a reunion of old ex-scientists.! It is natural to aim for a high conference level.

Thus, the submission phase started, and after this phase we received 27 submissions. Our 26 international Program Committee members then reviewed these submissions with one paper being reviewed by three different reviewers. As a result, 16 papers were accepted. Then, based on these accepted papers, the editorial board discussed whether the conditions had been met or not. We had to wait for about 20 days, until the editorial board finally sent us the good news of the acceptance of the proceedings as a CCIS volume. It was with the comment, "We expect and trust that you have arranged a thorough and competitive paper reviewing and selection process to ensure that only high-quality papers are accepted for publication that present substantial novel research achievements." Also, it was mentioned "the editorial board has noticed that the acceptance rate is quite above average for CCIS volumes. ... Still, in the future, we expect ICNNAI to increase the number of submissions, the geography of authors, and aim at decreasing the acceptance rate." We expect that, too. Now we are sure that ICNNAI 2014 was the landmark event in the history of this conference.

ICNNAI 2014 was made up of four categories of sessions: a plenary session, a regular session, a round table discussion, and a session with papers accepted after the deadline. All the sessions were held in a single track, and we offered enough time for hot discussions after each of the talks.

In the plenary session, the topic on forest maintenance was discussed. In Belarus, there is a vast forest called Belovezhskaya Pushcha National Park which is said to be the home of 900 plants and 250 animals and birds, including several rare species. As such, we invited Professor Koji Nakamura (Kanazawa University, Japan) whose specialty is maintenance of forest resources. The talk followed two short presentations about "Forests in Belarus" and "Forest Management from IT Prospects." The round table discussion started with a 40-minute talk on "Formal Definitions of Machine Intelligence." Then we discussed "whether we can achieve a human-like artificial intelligence using neural networks?" The session with papers accepted after the deadline was organized for new interesting results obtained after the deadline. Position papers or reports with premature results were also welcomed. The reports are not included in this volume but appear in a booklet published by the university press, which was delivered to all the participants.

Today there are numerous conferences on neural networks or on artificial intelligence. As far as we know, however, there exist is such conference on artificial intelligence specifically "by" neural networks. At this point in our conference series, papers that report on artificial intelligence "by" neural network are few, but as most of the papers are on neural network "or" artificial intelligence. We hope this conference will evolve to be a unique event where we can discuss the realizations of human-like intelligence "by" using neural networks in the foreseeable future.

June 2014   Vladimir Golovko
                                                        Akira Imada

Organization

ICNNAI 2014 was organized by the Department of Intelligent Information Technology, Brest Technical University.

Executive Committee

Honorary Chair

Petr Poyta Brest State Technical University

Conference Chair

Vladimir Golovko Brest State Technical University
Akira Imada Brest State Technical University

Conference Co-chairs

Vladimir Rubanov Brest State Technical University
Rauf Sadykhov Belarusian State University of Informatics and
Radioelecrtonics

Advisory Board

Vladimir Golenkov Belarusian State University of Informatics and
Radioelecrtonics
Valery Raketsky Brest State Technical University
Vitaly Sevelenkov Brest State Technical University

Program Committee

Dmitry Bagayev Kovrov State Techhological Academy, Russia
Irina Bausova University of Latvia, Latvia
Alexander Doudkin National Academy of Sciences of Belarus,
Belarus
Nistor Grozavu Paris 13 University, France
Marifi Guler Eastern Mediterranean University, Turkey
Stanislaw Jankowski Warsaw University of Technology, Poland
Viktor Krasnoproshin Belarusian State University, Belarus
Bora I. Kumova Izmir Institute of Technology, Turkey
Kurosh Madani University Paris-Est Creteil, France

Poramate Manoonpong	University of Southern Denmark, Denmark
Saulius Maskeliunas	VU Matematikos ir informatikos institutas, Lithuania
Jean-Jacques Mariage	Paris 8 University, France
Helmut A. Mayer	University of Salzburg, Austria
Alfonsas Misevicius	Kaunas University of Technology, Lithuania
Tsuneo Nakanishi	Fukuoka University, Japan
Jakub Nalepa	Silesian University of Technology, Poland
Vincenzo Piuri	Università degli Studi di Milano, Italy
Vladimir Redko	Russian Academy of Science, Russia
Izabela Rejer	West Pomeranian University of Technology, Poland
Hubert Roth	Universität Siegen, Germany
Anatoly Sachenko	Ternopil National Economic University, Ukrane
Sevil Sen	Hacettepe University, Turkey
Volodymyr Turchenko	University of Tennessee, USA
Lipo Wang	Nanyang Technological University, Singapore

Local Arrangements Committee

Alexander Brich
Andrew Dunets
Uladzimir Dziomin
Anton Kabysh
Valery Kasianik
Pavel Kochurko
Hennadzy Vaitsekhovich
Leanid Vaitsekhovich (Brest State Technical University)

Sponsoring Institutions

ERICPOL Brest (Dzerzhynskogo 52, 224030 Brest Belarus)

Fig. 1.

Table of Contents

Real World Application

Living in Harmony with Nature: Forest and "Satoyama" in Japan

Koji Nakamura

Kanazawa University
Kakuma Kanazawa Japan
kojink@staff.kanazawa-u.ac.jp

Abstract. In this article, we describe a project about forest from an ecological point of view where we argue a perspective of its current conditions, threats, challenges and future in the globe and Japan, together with our activities.

1 Forest and Its Multiple Functions

About 30% on average of the surface of the earth is covered with forests. The types of the forests are variable depending on the ecological conditions of habitats such as temperature, amount of rainfall, topography and soil. Forest has multiple functions beneficial to human society, which is called as ecosystem services (Millennium Ecosystem Assessment, carried out by UN in 2002-2005). Ecosystem services include provisioning (wood, charcoal, game animals, medical plants, mushroom and other foods), regulating (climate regulation, water purification, CO_2 absorption, preventing soil erosion and flood regulation), cultural (traditional festival and recreation) and supporting (nutrient cycling, soil formation and primary production). Biodiversity, fostered by forests, has crucially important roles for these functions.

As the results of Millennium Ecosystem Assessment indicated, forest degradation and fragmentation have been very rapidly increasing during last 50 years, which may have contributed to substantial net gains in human well-being and economic development, but these gains have been achieved at growing costs in the form of the degradation of many ecosystem services and increased risks of nonlinear changes.

2 Nature and Forest in Japan

Consisting of thousands of islands that vary greatly in size, Japan is a long archipelago stretching for approximately 3000 km from south to north located in the mid latitude (20-40 degrees) of the northern hemisphere. It expands from the subtropics to sub-frigid zones, so wide range of ecosystems have evolved. It has a complex topography from seacoast to mountain ranges with considerable differences in elevation (0-3779 m) and the four seasons are clearly defined due

V. Golovko and A. Imada (Eds.): ICNNAI 2014, CCIS 440, pp. 1–4, 2014.

to the effects of the monsoon climate. All of these factors combined have created diverse habitats and environments for the growth of plants and animals. At present, more than 90000 species have been confirmed as existing in Japan. Compared to other developed countries, Japan has an extremely high proportion of endemic species. In addition, extensive areas, where unique habitats have been created by humans.

Satoyama is a Japanese term applied to the border zone or area between mountain foothills and arable flat land. Literally, sato means arable and livable land or homeland, and yama means hill or mountain. Satoyama have been developed through centuries of small scale agricultural and forestry use.[1] In short, Satoyama is rural landscapes formed by sustainable agriculture and forestry.

Satoyama areas have also contributed to the country's rich biodiversity (See 'The National Biodiversity Strategy of Japan 2012-2020' Ministry of Environment of Japan, 2012).

Diversified forest types are found in Japan, growing in the diversified habitats and environments, as mentioned above. Forests cover about 66% of Japans land, which is the third largest forest coverage rate in the world, following Finland (72%, top) and Sweden (68%, second). However, in recent years, Japan has imported about 80% of woods consumed in the country from abroad due to a cheaper price, and forestry in Japan has been in economic slump for a long time. (See, also the next session.)

3   "Satoyama" in Japan and Its Global Significance

About 40-50% of Japan, ranging from lowland up to 500-800 m elevation, belonged to the Satoyama landscape, which was a mosaic of woodlands, plantations, grasslands, farmlands, irrigated ponds, canals, etc. In Japan, much attention has been drawn to the traditional Satoyama rural landscape because of its destruction and deterioration due to societal changes since the end of World War II.

In response to these trends, the Satoyama and Satoumi (coastal areas used for fishery, aquaculture and so on) Assessment (JSSA) was carried out as an international project, an assessment of the current state of knowledge of Satoyama and Satoumi, from 2007 to 2010, led by the Institute of Advanced Studies of United Nations University (UNU–IAS) and the Japanese Ministry of the Environment (Duraippah et al. 2012). As an international context, JSSA defines Satoyama landscapes as "dynamic mosaics of managed socio-ecological systems" producing a bundle of ecosystem services for human well-being, i.e., "socio-ecological production landscapes (SEPLS)."

The Satoyama Initiative (SI) started to promote (1) sustainable use of forest, agricultural land, pastoral land and other types of SEPLS, (2) Enhancing the resilience of SEPLS, (3) Valuing cultural and historical SEPLS and (4) strengthening multi-stakeholder partnership. The SI has used SEPLS as a key concept.

---

[1] From http://en.wikipedia.org/wiki/Satoyama.

It has become clear that these SEPLS and the sustainable practices and information they represent are increasingly threatened in many parts of the world. Commonly recognized causes include urbanization, industrialization, and rapidly shrinking rural populations. The SI has taken a global perspective and sought to consolidate expertise from around the world regarding the sustainable use of resources in SEPLS.

A global initiative relevant to the SI is "Globally Important Agricultural Heritage Systems (GIAHS)," which was launched by the Food and Agriculture Organization (FAO) of the United Nations in 2002. The overall goal of the initiative is to identify and safeguard GIAHS and their associated landscapes, agricultural biodiversity, and knowledge systems through catalyzing and establishing a long-term support program and enhance global, national, and local benefit derived through sustainable management and enhanced viability.

Traditional agriculture systems are still providing food for some two thousand million people in the world today, and also sustain biodiversity, livelihoods, practical knowledge, and culture. So far, a total of 25 GIAHS areas have been designated from 11 countries, including Algeria, Chile, China, India, Japan, Mexico, Morocco, Peru, Philippines, and so on (Berglund et al. 2014). See Fig. 1.

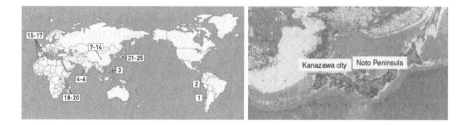

Fig. 1. The location of 25 GIAHS areas in the globe and in Japan

4   Current Project and a Future Challenge

In this section, we describe our activities to challenge against the current problems in Satoyama and Satoumi, i.e., the socio-ecological production landscape and seascape (SEPLS) of Noto Peninsula in Ishikawa Prefecture, located in Japan. The sea side of central Japan are rich in traditional culture, biodiversity and beautiful landscapes. Consequently, Noto's Satoyama and Satoumi was designated in June 2011 as Globally Important Agricultural Heritage Systems (GIAHS).

Today the ecosystems of these landscapes are under severe threats from various environmental and social problems, including under-management of the land, a shrinking rural population, declining agriculture and an ageing society, which result in the deterioration of ecosystem services and biodiversity as discussed in Japan Satoyama and Satoumi Assessment (JSSA, 2010).

he Noto Satoyama Meister Training Program of Kanazawa University (2007-2011) and its subsequent program (2012-) have succeeded in the capacity building of younger generations in the reactivation of rural communities in Noto Peninsula.

Ifugao Rice Terraces (IRT), a GIAHS site, in the Philippines has also been threatened by lack of young farmers and predominance of unregulated tourism activities. Therefore, it is an urgent need to develop local human resources for the sustainable development of IRT. We have just started the Ifugao Satoyama Meister Training Program, which aims to replicate the Satoyama Meister model in GIAHS IRT, in collaboration with Kanazawa University, University of the Philippines Open University and Ifugao State University with the support of the local government under the framework of JICA's Technical Cooperation for Grassroots Projects.

References

Duraiappah, A.K., Nakamura, K., Takeuchi, K., Watanabe, M., Nishi, M. (eds.): Satoyama-Satoumi Ecosystems and Human Well-being: Socio-ecological Production Landscapes of Japan. United Nations University Press (2012)

Berglund, B.E., Kitagawa, J., Lagerås, P., Nakamura, K., Sasaki, N., Yasuda, Y.: Traditional Farming Landscapes for Sustainable Living in Scandinavia and Japan: Global Revival Through the Satoyama Initiative. AMBIO (2014), doi:10.1007/s13280-014-0499-6

Ecological and Economical Aspects of Felling with Use of Multi-operational Felling Machines and Machinery

Marina V. Levkovskaya[1] and Vladimir V. Sarnatsky[2]

[1] Brest State University
Brest, Belarus
[2] Institute of Experimental Botany
Minsk, Belarus

Abstract. In this article, we consider influences of forestry equipments on pine tree plantation from a view point of ecology and economy. We investigate damages on trees after harvester felling, that is, inclination of growing trees, stripping barks, cutting branches and trunks and scrapping fallen leaves and branches. The object is to maintain clean pineries.

1   Introduction

Due to growing volume of mechanical cleaning cutting in middle-aged and maturing forests of Belarus, special topicality endowed to increase its profitability and quality by improving management and technology of felling, increasing level of mechanisation, reduction of damages left on tree's roots and rational log bucking. In places where machinery works there happens micro-climate changes of woodcutting area, soil compaction and physical properties of soil changes, under-story, growing trees, forest stand, that left, grassy and mossy coverage are damaged mechanically, that leads to corruption of surface soil or totally destroys individual components of it.

In order to study the influence of forestry equipment on pine plantations in pine forests of Baranochy forestry of Brest State Forestry Production Associations, 20 sample areas were upset. Objects of research were clean and mixed pineries of Baranochy's forestry, in which were made mechanical cleaning cutting. Duration of time after felling variates from 1 to 14 years, that allows to trace dynamics of changes in components of phytocenosis.

We identified next categories of visible damages on tree trunk after harvester felling: inclination of growing trees; bark stripping; scrapping branches; breaking and scrapping bark, trunk and branches; cuts of trunk. Share of damaged pine trunks in sample areas 5-10 years after mechanical cleaning cutting is 1.5-3 times lower, than in lumbered trees of this year cut, that could be explained with overgrowing of damages on trunk.

V. Golovko and A. Imada (Eds.): ICNNAI 2014, CCIS 440, pp. 5–8, 2014.
© Springer International Publishing Switzerland 2014

2 Analysis

From analysing mechanical felling of this year it was concluded that the lowest defectiveness of pine trees (4.5%) observed after felling in winter time. In spring-summer time (April-May) in comparison to winter, intensity of damage of roots and trunk are greater in 1.2-2 times. In spring-summer time when strength of bark is minimal there is chance of bark stripping. Low percentage of damaged trees is noted when cleaning cutting made by linear block technology. On trees, contiguous to skidder track at time of lumbering, share of damages sized below 100 cm^2 is about the half. Most of damages on sample areas of assortment logging is a bark stripping on height below 2.5 meters. In all cases most of damaged trees are concentrated on the border of technological corridors. Damages inflicted to tree's trunk occurs in time of felling trees by harvester and loading assortment. If felling made in winter time frozen ground and show protects roots and butt of trees from damages.

It is determined, that defectiveness of Scots pine (Pinus sylvestrisL) after felling in average is 5.9%, Norway spruce (Picea abies Karst) - 2.3%, silver birch (Betula pendula Roth) - 0.7%, that corresponds to technical requirements for preservation of the trunk and besides this damages do not lead to cessation of growth and desiccation of tree.

Acidity of the upper soil layers (pH) in the block of researched sample areas variates between 4.64 and 5.13, in the skidder track - from 4.74 to 5.33. In the felling area acidity of the soil reduced on 0.1-0.5 points and depends from kinds of growing plants.

For comparative analysis of influence of machinery on the density, hardness, moisture of the soil on sample area in zones of technological corridors and blocks samples of undisturbed soil were get from upper horizons (50 cm). Density and moisture of soil were determined in laboratory condition.

It is found, that transport trees additionally compacts the soil in skidder tracks in average on 6%, maximal value of soil compaction reaches 20%. Transport trees, in some cases, leads to soil compaction in forest in 1.1–1.2 times. If lumbering made in spring-summer time in pine forests maximal density of soil in run reaches 1.47 g/cm^3 and it exceeds limit in 1.2–1.4 times. In summer time soil density in skidder tracks is higher on 2-33%. Over time difference of soil density in skidder track and blocks reduces. So, if after felling it reaches 19%, then in places, where felling was in 2004, 2005 years in winter time it varies in limits of 2–7%.

Relatively more favourable conditions are in lumbering areas aged 8–9 years. Density of soil is reduced, that shows reversibility of the process of soil compaction. Changes of hardness of humus horizon are determined. Increase of hardness of soil under influence of transport trees are observed up to 10–17 kg/cm2 (in 2-4 times). Hardness of soil in skidder tracks exceeds on 37-73% the values for blocks, that is in 1.6-3.3 times higher. Moisture of soil in skidder tracks is lower, than in blocks, but higher than in test woodland. Increase of rainfall in skidder tracks, that reaches soil level is connected with removing of leaf canopy. Soil compaction in skidder tracks leads to reduction of its porosity, infiltration of water, changes in moisture regime and difficulties in water penetration. In some cases the reason of differences in moisture is reduction of infiltration of water in soils of corridors. Damage degree of soil mantle by forest machines in pine forest after felling is low in common or rarely medium.

3 Estimation

If its identified, that placing felling tailings on the skidder tracks in the process of lumbering significantly reduces negative influence of transport trees on the soil and decrease frost line. Severity of exposure to vegetable ground cover from logging machinery depends on season of felling, design features of machines and technological processes of cutting area operations. Regeneration and growth of forest range greatly depends on soil density, which also determines temperature, moisture, atmosphere conditions of soil and intensity of physico–chemical and biological processes, that are performed in soil. Changes of hydrophysical conditions of soil in 10 from 20 sample areas are observed.

After mechanical cleaning cutting mass of bioactive roots in soil of skidder tracks are decreased, that is connected with degradation of hydrophysical conditions of soil of skidder tracks. Dependence of level of root occupation in technological corridor and block could be traced.

Estimation of difference in level of root occupation in technological corridor and block are made by formula:

$$P = 100(m_c - m_b)/m_b(\%) \tag{1}$$

where m_c is mass of roots in technological corridor and m_b is mass of roots in block.

In first years after felling, when amount of trees are decreased, root occupation in soil are decreased, that helps remaining trees to increase feeding area. Thinning of forest stimulates growth of small roots. Root occupation in soil in thinned out and not forests gradually equalized and founded results are in agreement with other sources. Mass of roots in corridor and block were compared separately for soil horizons. Research results showed that in surface soil in technological corridor there is lower amount of roots in 1.2–2.2 times than in blocks and control areas, not depending on long standing of felling. This happens because surface soil in such situation are very prone to negative impact from felling machinery.

4 Summary

In time of thinning and accretion cutting defectiveness of trees are variates from 2% to 12.3%. In all cases most of damaged trees are concentrated on border of technological corridors. Hydrophysical conditions of soil of skidder tracks and blocks are subjected to considerable changes depending on technology and age of felling, initial differences in physical characteristics of soil and on season in which clearing cutting was made.

Making the felling in winter time, when soil is frozen and there is snow coverage, has a positive impact on state of soil. In connection with soil compaction after mechanical clearing cutting in the soil of skidder tracks reduced mass of small roots of pine.

Degradation of hydrophysical conditions of soil increases time for regeneration of mass of roots. It makes sense to continue researches on this objects for determining the duration of influence of lumbering machinery on different components of forests.

With a purpose of increase of economical efficiency of harvesting of wood and use of lumbering machinery it its necessary to organize so called concentrated felling, that held in limits of not big forest area or 1–3 blocks including clearing cutting, sanitary felling, final felling and other kinds of felling. Important technological aspect is a reduce of non-productive loses in work of machines and machinery by reducing run unload and distance of dragging, skidding of converted wood, transportation of it to the storage area.

(The original draft was written in Russian. Aleksandr Brich (Brest State Technical University) translated the draft into English.)

A Literature Review: Forest Management with Neural Network and Artificial Intelligence

Akira Imada

Department of Intelligent Information Technology
Brest State Technical University
Moskowskaja 267, Brest 224017 Belarus
`akira-i@brest-state-tech-univ.org`

Abstract. The 8th International Conference on Neural Network and Artificial Intelligence invites a plenary talk whose topic is *Forest Resource Maintenance*. It's not specifically from the point of information technology but from a general point of view. Thinking of the title of this conference, *'neural networks and artificial intelligence,'* the talk follows this literature review on the topic of *forest management by neural network and/or artificial intelligence*. While focus is mainly put on *wildfire prediction*, also *preservation of biodiversity of forest ecosystems* and *forest resource management* are surveyed. Comparison of these methods with traditional statistical methods such as regression is mentioned too.

1   Introduction

As catastrophe usually comes all of a sudden unpredictably, people used to rely on fortune telling in our history to avoid such catastrophes, or to reduce calamities and tragedy caused by them. Hence, fortune telling also played an important role in theaters, such as Victor Herbert's operetta 'Fortune Teller,' 'Gypsy Baron' by Johan Strauss, or not to mention but also Bizet's 'Carmen.' Safi (2013) wrote, *"As predicting what might happen in the future has always been considered as a mysterious activity, scientists in modern era have tried to turn it into a scientific activity based on well-established theories and mathematical models."*

We now take a look at those approaches in the literature of how we predict what is going to happen in a forest such as wildfire, biodiversity, resources or something else, in which a well-established neural network model is exploited, or artificial intelligence is claimed to make it.

Selection of literature's is not optimized but rather spontaneous, expecting this article to be a set of initial pointers for readers' own survey.

2   Using Artificial Neural Network

Peng (1999) wrote, *"Data concerning forest environment are sometimes obscure and unpredictable, artificial neural network, which is good at processing such*

V. Golovko and A. Imada (Eds.): ICNNAI 2014, CCIS 440, pp. 9–21, 2014.
© Springer International Publishing Switzerland 2014

a non-linearity, has been extensively explored since late 1990's as an alternative approach to the classical method of modeling complex phenomena in forest (McRoberts et al. 1991; Gimblett et al. 1995; Lek et al. 1996; Atkinson et al. 1997)." We see such approaches in this section.

2.1 Forest Wildfire Prediction

Every year we hear quite a lot of news of wildfire somewhere in the globe, such as in US, Turkey, Greece, Spain, Lebanon and so on and on. Sometimes wildfire kills people or even firefighters. The article in New York times on 30 June 2013 reads:

> *Nineteen firefighters were killed on Sunday battling a fast-moving wildfire menacing a small town in central Arizona. The firefighters died fighting the Yarnell Hill Fire near the town of Yarnell, about 80 miles northwest of Phoenix. ... There were several fires still active in the Yarnell area. In a search of the scene, crews found the bodies of the firefighters.*

Let's see how frequently wildfires had happened in U.S. in 2013, as an example. Tres Lagunas fire, Thompson Ridge fire, Silver fire, Jaroso fire in New Mexico; Black Forest fire, Royal Gorge fire in Colorado; Yarnell Hill fire in Arizona (where 19 firefighters were killed as cited above); Quebec fire in Quebec; Mount Charleston fire, Bison fire in Nevada; Idaho Little Queens fire in Idaho; Silver fire, Beaver Creek fire, Rim fire, Morgan fire, Clover fire in California.[1]

Vasilakos et al. (2009) estimated, the percentage of the influence of lots of factors to fire ignition risk in Lesvos Island in Greece. Here let's see, at first, his well organized survey on wildfire prediction in detail, which would help us make a further survey in this topic.

Vasilakos wrote, *"Wildland fire danger evaluation is an integration of weather, topography, vegetative fuel, and socioeconomic input variables to produce numeric indices of fire potential outputs (Andrews et al. 2003; Pyne et al. 1996)."* He went on, *"Various quantitative methods have been explored for the correlation of the input variables in fire danger assessment; most of these methods include the input variables' importance as a direct or indirect output."* Traditionally, statistical methods were widely used for fire danger calculation. Vasilakos further wrote, *"More specifically, linear and logistic regression techniques were proposed, so the coefficients of the models reflect the influence of inputs on fire danger (Chou 1992; Chou et al. 1993; Kalabokidis et al. 2007; Vasconcelos et al. 2001).*

"Artificial neural networks have been also used in fire ignition risk estimation (Chuvieco et al. 1999; Vasconcelos et al. 2001; Vasilakos et al. 2007; Vega-Garcia et al. 1996). ... In our previously published research (Vasilakos et al., 2007), three different neural networks were developed and trained to calculate three intermediate outcomes of Fire Ignition Index, i.e., the Fire Weather Index, the Fire Hazard Index, and the Fire Risk Index."

---

[1] Extracted from `http://en.wikipedia.org/wiki/List_of_wildfires`.

Multilayer Perceptron is an architecture of the neural network that most such reports exploit. Fig. 1 is the one of those multilayer perceptorns Vasilakos used. And the backpropagation algorithm[2] is used for training of these neural networks in most of such approaches.

Vasilakos (2009) analyzed wildfire data in the history of Lesvos Island, and then designed a neural network that was trained with using such previous data. The neural network receives a set of inputs that may influence the output that shows us a probability of the risk of occurrence of the wildfire. One important question is, how can we know the degree of importance of those factors given as the inputs.

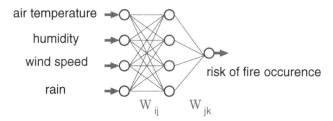

Fig. 1. One example of Multilatyer perceptron from those used by Vasilakos (2009) to know which input is the most influential factor for the output of ignition of fire occurrence

Logistic Regression is a well established method to estimate the probability of a binary independent variable from a set of independent continuous variables. It can be used to know the degree to how important the influences of dependent continuous variables $x_1, x_2, \cdots$ to the binary independent variable y.

To imagine how it works, let's now consider a simple situation where provability y depends only on one variable x. Then sample data of (x, y) could be assumed to fulfill the equation

$$y = \frac{1}{1 + exp\{-(a_0 + a_1 x)\}} \tag{1}$$

where two parameters a_0 and a_1 could be estimated so that these sample points are best fit to the graph of this equation. For the purpose, we can use, for example, maximum likelihood estimation. Then we can infer the probability y of any x given.

Now we have n independent variables $x_1, x_2, \cdots, x_n$. Then dependent variable y is expressed as n-dimensional logistic function

$$y = \frac{1}{1 + exp\{-(a_0 + a_1 x_1 + a_2 x_2 + \cdots + a_n x_n)\}} \tag{2}$$

---

[2] In this paper, assuming all the readers are familiar with the basic idea of neural network, allow us not to refer to who firstly proposed or where we can obtain a detailed idea for the well known methodology such as *backpropagation*.

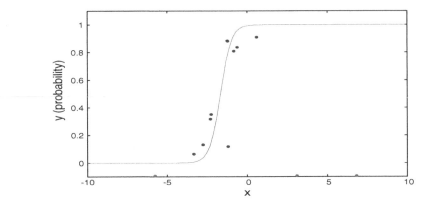

Fig. 2. A fictitious example of logistic regression from nine different values of x, after adjusting two parameters a_0, and a_1 by the maximum likelihood estimation

where a_i $(i = 0, 1, \cdots, n)$ are parameters. After adjusting these $(n+1)$ parameters such that the points in n-D space corresponding to the sample data given fit this hyper surface with maximum likelihood, we can interpret a_i as the degree of influence of x_i to y.

Garson's Method is an algorithm to measure the relative importance of input variables of an already successfully trained neural network based on its connection weights (Garson 1991). Assuming now w_{ij}'s are the connection weights between N input neurons and L hidden neurons, and u_{jk}'s are the connection weights between L hidden neurons and M output neurons, the percentage of influence Q_{ik} of input x_i on the output y_k is estimated by

$$Q_{ik} = \frac{\sum_{j=1}^{L} \left(\frac{w_{ij}}{W_j} u_{jk} \right)}{\sum_{i=1}^{N} \left(\sum_{j=1}^{L} \left(\frac{w_{ij}}{W_j} \right) \right)} \tag{3}$$

where

$$W_j = \sum_{r=1}^{N} w_{rj} \tag{4}$$

for normalization.

Using logistic-regression, Garson's equation and some other methods, Vasilakos et al. (2008) estimated, the percentage of the influence of lots of factors to fire ignition risk in Lesvos Island. Here let's see the result by Garson's equation, among others. The degree of importance of air temperature, wind speed, humidity and amount of rainfall to the risk of fire occurrence were found to be 28.7%, 20.9%, 14.5% and 35.9%, respectively, where in this example, the neural network was feedforward one with four input neurons, four hidden neurons and one output neuron trained by the backpropagation. The other factors chosen by the authors were altitude, distance to urban areas, day of the week, month of the year, etc. Thus, the authors determined influential ones out of 17 factors, with dependent variable being binary expressing presence or absence of fire ignition possibility. Training and validation samples were created from the total fire history database.

Support Vector Machine is, to simply put, a method to classify objects in a multi-dimensional space with hopefully by hyperplane, or otherwise a hyper surface. Sakr et al. (2010) proposed a forest fire risk prediction algorithm based on Support Vector Machines. The algorithm predicts the fire hazard level of the day from previous weather condition. The algorithm used the data from a forest in Lebanon for training.

Safi et al. (2013) used similar approach with forest fire data from the wild area of 700 square kilometers of ancient oak forests in the Portuguese Montesinho Natural Park with the output signal representing the total surface in hectare of the corresponding burned area.

2.2   Preservation of Biodiversity of Forest Ecosystems

Gil-Tena et al. (2010) modeled bird species richness in Catalonia, Spain. The authors wrote, *"Forest characteristics that determine forest bird distribution may also be influencing other forest living organisms since birds play a key functional role in forest ecosystems and are often considered good biodiversity indicators (Sekercioglu 2006)."*

Then authors exploited a three layer feedforward neural network with training being based on adaptive gradient learning, a variant of backpropagation. After optimizing the structure of neural network, estimated bird richness values by each neural network model are compared with the observed values in the real forest, and evaluated using linear correlation. Forest bird species richness was obtained using presence/absence data of 53 forest bird species as well as 11 data concerning forest (such as tree-species-diversity and past-burnt-area), five data on climate (such as temperature and precipitation) and 5 data on human pressure (such as road density and human population). Those bird data were collected by the volunteers from the Catalan Breeding Bird Atlas (Estrada et al. 2004). The other data are obtained from various sources, such as Spanish Forest Map, Forest Canopy Cover, Catalan Department Environment and Housing, Spanish Digital Elevation Model, National Center of Geographical Information, etc. (See references cited therein.) Data were divided into two groups. One was used for training and the other was for validation.

Also from this aspect, using neural networks, Peng et al. (1999) studied which of the forest features correlate with biodiversity, and then modeled forest bird species richness as a function of environment and forest structure. The authors wrote, *"Much progress has been made in this area since the initial use of artificial neural network to model individual tree mortality in 1991 (Guan and Gertner*

1991a). In the same year, Guan and Gertner (1991b) successfully developed a model, based on an artificial neural network, that predicts red pine tree survival."

To know this topic more in detail, The Ph.D dissertation by Fernandez (2008) might be good to be read.

2.3   Forest Cover Type Prediction

Forest cover type is a classification system based on trees that predominate in a particular area, as defined by Steve Nix.[3] Figure 3 suggests an image that shows the distribution of 25 classes of general forest cover such as (Eastern Oak_Pine forests) as well as water and non-forest land, in the United States and Puerto Rico.[4] When we concern a management of land, a 'Natural Resource Inventory' is a vital information. 'Forest cover type' is one of the most basic items in such inventories (Blackard et al. 1999). Blackard et al. (1999) predicted forest cover

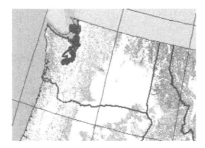

Fig. 3. An example of forest cover maps (taken from the web page by NationalAtras.gov)

types in the four wilderness areas of the Roosevelt National Forest in northern Colorado – Rawah, Comanche Peak, Neota, and Cache la Poudre. A feedforward neural network model was used. After looking for an optimal architecture by trial and error experiment, the neural network was made up of 54 input neurons, 120 hidden neurons, and 7 output neurons. Training was by backpropagation. Then they compared the results with the results by a traditional statistical model based on Gaussian discriminant analysis, and found a more accurate prediction by neural network.

In addition to the accuracy, Blackard et al. (1999) wrote, *"Recording the data by human is prohibitively time consuming and/or costly. ... Furthermore, an agency may find it useful to have inventory information for adjoining lands that are not directly under its control, where it is often economically or legally impossible to collect inventory data. Predictive models provide an alternative method for obtaining such data."*

---

[3] http://forestry.about.com/cs/glossary/g/for_cov_type.htm.
[4] http://nationalatlas.gov/mld/foresti.html.

Meyer (2001), then student of Wisconsin University, made a similar study in his term project.[5] He used a Support Vector Machine, with input being 54 variables, of which 10 were quantitative measures such as altitude, distance to water, while the remaining 44 were Boolean values representing soil conditions such as soil-type and wilderness-type. He classified these data into 7 classes, and the outputs showed one of these 7 classes. Almost 600,000 samples were used for training and testing. The data he used were forest cover type data set from the University of California-Irvine Knowledge Discovery in Databases Archive[6] compiled initially by Blackard (1998). Thus the Support Vector Machine classified the input data to one of the seven forest cover type with a success rate of 70%.

Peng et al. (1999) cited (Campbell et al. 1989) and (Downey et al. 1992) as studies in which neural network classifies land cover using data from Landsat satellite. Bennediktsson et al. (1990) used Landsat multispectral scanner network imagery and three topographic data sets (elevation, slope and aspect) to classify land cover. Peddle et al. (1994) applied the neural network approach to classify land cover in Alpine regions from multi-source remotely sensed data. Gong et al. (1996) have tested the feasibility of applying feedforward neural network and backpropagation to land system. Pattie et al. (1996) compared neural network model with regression model on forecasting wilderness.

2.4   Forest Resource Management

Peng et al. (1999) cited Coulson et al. (1987) as a study that had started to apply 'expert system' to forest resources management. The authors also cited (Gimblett et al. 1995; Lek et al. 1996) as an emergence of artificial neural network model as an alternative approach for modeling nonlinear and complex phenomena in forest resources management. Let's see such studies now.

Growth Model. Castro (2013) proposed a model of the growth of eucalyptus in northern Brazil, to predict tree height, diameter and mortality probability by neural network. Lots of architectures of neural network were trained using real data from the forest with the purpose being looking for the optimized architecture. Then, the author concluded that neural network may work as an alternative to the traditional procedure such as regression analysis with linear or nonlinear functions, to model an individual tree. Furthermore, an artificial neural network may help identify the most critical input variables to predict diameter and height growth as well as mortality probability, and provide a better understanding of the dynamic of models at the individual tree level, becoming a valuable tool for eucalyptus forest management.

Castro also cited (Merkl et al. 1998), (Weingartner et al. 2000) and (Hasenauer et al. 2001) as a study of mortality prediction by neural network, (Diamantopoulou 2005) as a study of tree volume estimation also by neural network,

---

[5] http://homepages.cae.wisc.edu/~ece539/project/f01/meyer.pdf .
[6] http://kdd.ics.uci.edu .

(Paruelo et al. 1997), (Gevrey 2003) and (Leite et al. 2011) as a report of estimation of forest resources by regression.

Prediction of Tree Mortality. In order to predict individual tree mortality, Weingartener et al. (2000) compared multi layer perceptron, Learning Vector Quantization, and Cascade Correlation Networks with the conventional model using logit function (i.e., inverse of the sigmoidal logistic function). Training these three networks with the data from the Austrian National Forest Inventory, they compared the performance of each networks of mortality prediction using dataset from the Litschau Forest, and concluded Learning Vector Quantization slightly outperformed the others.

Common Pool Resource is a resource that benefits a group of people, but which provides diminished benefits to everyone if each individual pursues his or her own self interest, such as forest river.

Ulrich (2013) wondered why some communities succeed in managing common pool resources, while others fail. By restricting his analysis only to traditional commons of land use, forest management, irrigation and fisheries, he tried to look for essential factors to explain why. He cited (Hess 2008) as a more comprehensive overview of common pool problems. He wrote, *"We don't have universal such success factors that can explain any system concerning common pool resource (Meinzen-Dick 2007). ... Some comprehensive set of such success factors have already reported (See, e.g., (Pagdee et al. 2006)) which comprises more than 100 factors. Or there are consensus on a set of 20 to 30 success factors."*

Ulrich (2013) sought to find a set including less factors but a result as comprehensive as possible, starting with such reports as (Ostrom 2009) and (Agrawal 2001). Ulrich briefly exemplified the methodology using data on Nepal irrigation systems collected in the 'Nepal Irrigation Institutions and Systems' database.[7] It contains 263 cases with 478 variables per case. The cases were coded during 1982 and 1997. He continued, *"For further information see (Tang 1989)."*

3 Using Artificial Intelligence

It's not difficult to search for papers that claim artificial intelligence. There exist lots of such papers that propose a new method for predicting a future. Let's take a look at those already published papers, for example, on wildfire prediction, the title of which includes the term *'artificial intelligence.'*

In their paper, Peng et al. (1999) wrote, *"The application of artificial intelligence in forest and natural resources management started with the development of expert systems (Coulson et al. 1987)."*

---

[7] NIIS research team (1993) "Nepal irrigation institutions and systems' database coding sheets and forms." Indiana University, Workshop in political theory and policy analysis.

Since then, indeed, not a few approach have claimed that they use the artificial intelligent methodology. Let's name a few. Kourtz (1990) studied forest management in all aspects of Canadian forestry by expert system in his paper entitled *'Artificial intelligence: a new tool for forest management.'* Arrue et al. (2000) proposed a system to detect forest fires by using computer vision, neural networks and expert fuzzy rules, in their paper entitled *'An intelligent system for false alarm reduction in infrared forest-fire detection.'* Actually, the late 1990's was a dawn of artificial intelligence and they dreamed a bright future of establishing a *human-like artificial intelligence*. But nowadays, we don't think the state of the art then had such a bright future.

Now let's see more recent ones. Angayarkkani et al. (2010) proposed a system of detecting forest fires in their paper *'An intelligent system for effective forest fire detection using spatial data.'* The digital image in the forest area were converted from RGB to XYZ color space, and then segmented by employing anisotropic diffusion to identify fire region. Radial basis function neural networks was employed.

The title of already mentioned paper by Sakr et al. (2010) was *'Artificial intelligence for forest fire prediction.'*

But Are Those Intelligence Really Intelligent? In his paper, Castro (2013) wrote, *"Artificial neural networks are a type of artificial intelligence system similar to human brain, having a computational capability which is acquired through learning."*

In fact, many claim their proposed machine to be intelligent. However, are these machines really intelligent like human intelligence as they claim? We cannot be so sure. For example, the paper by Sakr et al. (2010) reads, *"The methods are based on artificial intelligence"* in the abstract, while the term *'artificial intelligence'* never appeared afterwords in the whole text. Instead, he concluded, just *"advanced information communication technologies could be used to improve wildfire prevention and protection."* That's all there is to it.

Yet another such example is the paper by Wendt et al. (2011) entitled *'Input parameter calibration in forest fire spread prediction: Taking the intelligent way.'* The appearance of the term *'intelligent'* is only once, i.e., *"Evolutionary Intelligent System"* without mentioning what is that.

Intelligent Robot Might do Human Dangerous Jobs. Firefighters' jobs are crucially dangerous. The same goes, more or less, for other jobs such as policemen, soldiers, astronauts etc. It will be nice if we can replace them with intelligent robots. As a human-like intelligence is sometimes required for such robots, this could be one of our strong motivations to develop machine intelligence.

A recent article in the New York times[8] reads:

---

[8] From the article with the headline "Border's new sentinels are robots, penetrating deepest drug routes," in the New York Times on-line on 22 February 2014.

Tom Pittman has made a career as a Border Patrol agent here guarding this city's underground drainage system, where the tunnels that carry sewage and storm runoff between the United States and Mexico are also busy drug-smuggling routes. Over the years, he has crawled and slithered past putrid puddles, makeshift latrines and discarded needles left behind by drug users, relying on instincts, mostly, to gauge the risks ahead. It is a dirty and dangerous business, but these days, there is a robot for that.

As the article went on *"The robots can serve as the first eyes on places considered too risky for humans to explore,"* the aim is not a creation of intelligent robot agent, at least up to this moment. However, there would be a evolutionary race between the smugglers and robots. Which will become more intelligent next time?

Toward More Human-Like Intelligence. Stipanicev (2011) proposed what he calls the 'Intelligent Forest Fire Monitoring System - iForestFire' in his paper whose title is, *'Intelligent forest fire monitoring system - from idea to realization.'* The goal is *"to achieve early wildfire detection, which is traditionally based on human wildfire surveillance, realized by 24 hours observation by human located on selected monitoring spots in Dalmatian forest in Croatia which belongs to countries with high wildfire risk,"* he wrote. *"With its ultimate aim being replacing human with intelligent machine, operators could be located on any location with broadband Internet connection and its user interface to the standard Web browser.*

"The system is an example of 'Future Generation Communication Environment' where all applications and services are focused on users, and 'user' in our case is the natural environment, having the main task – wildfire protection. For such environment behavior the term 'environmental intelligence' was introduced (Stipanicev et al. 2007; Seric et al. 2009)."

Thus, the author claims *"iForestFire is intelligent because it is based on artificial intelligence, computational intelligence and distributed intelligence technologies such as multi-agent based architecture where all agents share the same ontology and speak the same agent communication language."*

4   Concluding Remarks

To cover the plenary talk on 'forest resource maintenance' in general, a literature survey on the topics specifically from IT point of view has made, with focus being *wildfire prediction, preservation of biodiversity in ecosystems, forest cover type prediction, forest resource maintenance* such as *common pool resource.*

In Belarus, we have a huge forest called *'Belovezhskaya Pushcha National Park'* where it is said to be *"the home to 900 plants and 250 animals and birds, including several rare species."* Hence, contributions to a maintenance of this ecological environment is a duty to us IT scientists in Belarus. Further, this issue is going to be worldwide now in this era of global warming. The author wish this small survey paper to play a role of useful pointers for this field.

References

Agrawal, A., et al.: Explaining success on the commons: community forest governance in the Indian Himalaya. World Development 34(1), 149–166 (2006)

Andrews, P.L., et al.: BehavePlus fire modeling system user's guide, v. 2.0. General technical report RMRS-GTR-106WWW, USDA, Forest Service, Rocky Mountain Research Station (2003)

Angayarkkani, K., et al.: An intelligent system for effective forest fire detection using spatial data. International Journal of Computer Science and Information Security 7(1) (2010)

Arrue, B.C.: An intelligent system for false alarm reduction in infrared forest-fire detection. IEEE Intelligent Systems and their Applications 15(3), 64–73 (2000)

Atkinson, P.M., et al.: Introduction: Neural networks in remote sensing. International Journal of Remote Sensing 18, 699–709 (1997)

Benediktsson, J.A., et al.: Neural network approaches versus statistical methods in classification of multisource remote sensing data. IEEE Transaction on Geoscience and Remote Sensing 28, 540–552 (1990)

Blackard, J.A.: Comparison of neural networks and discriminant analysis in predicting forest cover types. Ph.D. dissertation, Department of Forest Sciences, Colorado State University (1998)

Blackard, J.A.: Comparative accuracies of artificial neural networks and discriminant analysis in predicting forest cover types from cartographic variables. Computers and Electronics in Agriculture 24, 131–151 (1999)

Braitenberg, V., et al.: Cortex: statistics and geometry of neuronal connectivity. Springer (1997)

Campbell, W.J., et al.: Automatic labeling and characterization of objects using artificial neural networks. Telematic and Informatics 6, 259–271 (1989)

Castro, R.V.O.: Individual growth model for eucalyptus stands in Brazil using artificial neural network. In: International Scholarly Research Network, ISRN Forestry Volume. Hindawi Publishing Corporation (2013)

Chou, Y.H.: Spatial autocorrelation and weighting functions in the distribution of wildland fires. International Journal Wildland Fire 2(4), 169–176 (1992)

Chou, Y.H., et al.: Mapping probability of fire occurrence in San Jacinto Mountains, California, USA. Environment Management 17(1), 129–140 (1993)

Coulson, R.N., et al.: Artificial intelligence and natural resource management. Science 237, 26–67 (1987)

Chuvieco, E., et al.: Integrated fire risk mapping. In: Remote Sensing of Large Wildfires in the European Mediterranean Basin. Springer (1999)

Diamantopoulou, M.J.: Artificial neural networks as an alternative tool in pine bark volume estimation. Computers and Electronics in Agriculture 48(3), 235–244 (2005)

Downey, I.D., et al.: A performance comparison of Landsat thematic mapper land cover classification based on neural network techniques and traditional maximum likelihood algorithms and minimum distance algorithms. In: Proceedings of the Annual Conference of the Remote Sensing Society, pp. 518–528 (1992)

Estrada, J., et al.: Atles dels ocells nidificants de Catalunya 1999–2002. In: Institut Catal d'Ornitologia (ICO)/Lynx Edicions, Barcelona, España (2004)

Fernandez, C.A.: Towards greater accuracy in individual-tree mortality regression. Ph.D dissertation, Michigan Technological University (2008)

Frey, U.J., et al.: Using artificial neural networks for the analysis of social-ecological systems. Ecology and Society 18(2) (2013)

Garson, G.D.: Interpreting neural-network connection weights. AI Expert Archive 6(4), 46–51 (1991)

Gevrey, M., et al.: Review and comparison of methods to study the contribution of variables in artificial neural network models. Ecological Modelling 160(3), 249–264 (2003)

Gil-Tena, A., et al.: Modeling bird species richness with neural networks for forest landscape management in NE Spain. Forest Systems 19(SI), 113–125 (2010)

Gimblett, R.H., et al.: Neural network architectures for monitoring and simulating changes in forest resources management. AI Applications 9, 103–123 (1995)

Gong, P., et al.: Mapping ecological land systems and classification uncertainties from digital elevation and forest-cover data using neural network. Photogrammetric Engineering and Remote Sensing 62, 1249–1260 (1996)

Guan, B.T., et al.: Using a parallel distributed processing system to model individual tree mortality. Forest Science 37, 871–885 (1991a)

Guan, B.T., et al.: Modeling red pine tree survival with an artificial neural network. Forest Science 37, 1429–1440 (1991b)

Hasenauer, H., et al.: Estimating tree mortality of Norway spruce stands with neural networks. Advances in Environmental Research 5(4), 405–414 (2001)

Hess, C.: Mapping the new commons. Governing shared resources: connecting local experience to global challenges. In: Proceedings of the Twelfth Biennial Conference of the International Association for the Study of Commons (2008)[9]

Kalabokidis, K.D., et al.: Geographic multivariate analysis of spatial fire occurrence. Geotechnical Scientific Issues 11(1), 37–47 (2000) (in Greek language)

Kalabokidis, K.D., et al.: Multivariate analysis of landscape wildfire dynamics in a Mediterranean ecosystem of Greece. Area 39(3), 392–402 (2007)

Kourtz, P.: Artificial intelligence: A new tool for forest management. Canadian Journal Forest Research 20, 428–437 (1990)

Leite, H.G.: Estimation of inside-bark diameter and heartwood diameter for Tectona Grandis Linn trees using artificial neural networks. European Journal of Forest Research 130(2), 263–269 (2011)

Lek, S., et al.: Application of neural networks to Modeling nonlinear relation-ships in ecology. Ecological Model. 90, 39–52 (1996)

McRoberts, R.E., et al.: Enhancing the Scientific process with artificial intelligence: Forest science applications. AI Applications 5, 5–26 (1991)

Meinzen-Dick, R.: Beyond panaceas in water institutions. Proceedings of the National Academy of Sciences of the United States of America 104(39), 15200–15205 (2007)

Merkl, D., et al.: Using neural networks to predict individual tree mortality. In: Proceedings of International Conference on Engineering Applications of Neural Networks, pp. 10–12 (1999)

Meyer, B.: Forest cover type prediction. Course ECE 539 Term Project, Wisconsin University (2001)

Ostrom, E.: A general framework for analyzing sustainability of social-ecological systems. Science 325, 419–422 (2009)

Paruelo, J.M.: Prediction of functional characteristics of ecosystems: A comparison of artificial neural networks and regression models. Ecological Modeling 98(2-3), 173–186 (1997)

---

[9] It was difficult to find the book of this volume any more, if not at all. So information of 'pp' cannot be shown here. The paper is available at
http://works.bepress.com/charlotte_hess/6.

Pattie, D.C., et al.: Forecasting wilderness recreation use: Neural network versus regression. AI Application 10(1), 67–74 (1996)

Peddle, D.R., et al.: Multisource image classification II: An empirical comparison of evidential reasoning, linear discriminant analysis, and maximum likelihood algorithms for alpine land cover classification. Canadian Journal Remote Sensing 20, 397–408 (1994)

Peng, C., et al.: Recent applications of artificial neural networks in forest resource management: An overview. In: Environmental Decision Support Systems and Artificial Intelligence, pp. 15–22 (1999)

Pyne, S.J., et al.: Introduction to wildland fire, 2nd edn. Wiley (1996)

Safi, Y., et al.: Prediction of forest fires using artificial neural networks. Applied Mathematical Sciences 7(6), 271–286 (2013)

Sakr, G.E., et al.: Artificial intelligence for forest fire prediction. In: Proceeding of International Conference on Advanced Intelligent Mechatronics, pp. 1311–1316 (2010)

Sekercioglu, C.H.: Increasing awareness of avian ecological function. Trends of Ecological Evolution 21, 464–471 (2006)

Seric, L., et al.: Observer network and forest fire detection. Information Fusion 12, 160–175 (2011)

Stipanicev, D.: Intelligent forest fire monitoring system - from idea to realization. In: Annual 2010/2011 of the Croatian Academy of Engineering (2011)

Tang, S.Y.: Institutions and collective action in irrigation systems. Dissertation. Indiana University (1989)

Ulrich, J.F., et al.: Using artificial neural networks for the analysis of social-ecological systems. Ecology and Society 18(2), 42–52 (2013)

Vasconcelos, M.J.P., et al.: Spatial prediction of fire ignition probabilities: Comparing logistic regression and neural networks. Photogramm Engineering Remote Sensors 67(1), 73–81 (2001)

Vasilakosi, C., et al.: Integrating new methods and tools in fire danger rating. International Journal of Wildland Fire 16(3), 306–316 (2007)

Vasilakosi, C., et al.: Identifying wildland fire ignition factors through sensitivity analysis of a neural network. Natural Hazards 50(1), 12–43 (2009)

Vega-Garcia, C., et al.: Applying neural network technology to human caused wildfire occurrence prediction. Artificial Intelligence Application 10(3), 9–18 (1996)

Weingartner, M., et al.: Improving tree mortality predictions of Norway Spruce Stands with neural networks. In: Proceedings of Symposium on Integration in Environmental Information Systems (2000)

Wendt, K., et al.: Input parameter calibration in forest fire spread prediction: Taking the intelligent way. In: Proceedings of the International Joint Conference on Artificial Intelligence, pp. 2862–2863 (2011)

Yoon, S.-H.: An intelligent automatic early detection system of forest fire smoke signatures using Gaussian mixture model. Journal Information Process System 9(4), 621–632 (2013)

Can Artiticial Neural Networks Evolve to be Intelligent Like Human?

A Survey on "Formal Definitions of Machine Intelligence"

Akira Imada

Department of Intelligent Information Technology
Brest State Technical University
Moskowskaja 267, Brest 224017 Belarus
`akira-i@brest-state-tech-univ.org`

Abstract. The 8th International Conference on Neural Network and Artificial Intelligence organizes a round table discussion where, thinking of the title of this conference *'neural network and artificial intelligence,'* we discuss whether we will be able to achieve a real human-like intelligence by using an artificial neural network, or not. This article is to break the ice of the session. We argue how these proposed machines, including those by neural networks, are intelligent, how we define machine intelligence, how can we measure it, how those measurements really represent an intelligence, and so on. For the purpose, we will take a brief look at a couple of formal definitions of machine intelligence so far proposed. We also take it a consideration on our own definition of machine intelligence.

1 Introduction

Since John McCarthy at MIT coined the term *"Artificial Intelligence"* in 1956 aiming to make a machine acquire a human-like intelligence in a visible future, we have had lots of discussions whether it is possible in a real sense, and actually lots of what they call an intelligent machine have been reported. The term is ubiquitous in our community these days. Let's name a few: *'intelligent route finding,'* *'intelligent city transportation system,'* *'intelligent forecasting stock market,'* etc. Then the question arises, "Degree to how they are intelligent, indeed?"

Or, are they not intelligent at all? Malkiel (2007) once wrote, *"A monkey throwing darts at the Wall Street Journal to select a portfolio might be better than the one carefully selected by experts"*

Lots of research works are now on going to understand the mechanism of brain using state of the art technology. For example, we can map human (or no-human) brain by fMRI, or we can measure micro-voltage fluctuation in any part of the brain by EEG. Thus, by capturing a snapshot of the brain in action, we can observe discrete activities that occur in specific locations of the brain. On the other hand, Weng (2013) wrote, *"A person who does not understand how a computer works got the wiring diagram of a computer and all snapshots*

V. Golovko and A. Imada (Eds.): ICNNAI 2014, CCIS 440, pp. 22–33, 2014.

of voltages at all the spots of the wires as a high definition fMRI movie. Then he said "now we have data, and we need to analyze it!" Our efforts are like *"Tree climbing with our eye's on the moon?"* as once Dreyfus (1979) sarcastically wrote?

A Legendary Chess Automaton in 18th Century. It had been a long time dream to create a machine which can play chess like a human. See, for example, the book by Standage (2002) that described an eighteenth century chess-playing machine. In 1770, Wolfgang von Kempelen created a machine called 'Turk,' and claimed it plays chess like human. It is said that his intention was to impress Maria Theresa, and infact, the machine debuted in Schönbrunn Palace in Vienna. Napoleon Bonaparte was one of those who played with this machine. Then the machine toured in Europe after 1783. In 1811, Märzel (also known as the inventor of metronome) became interested in this machine and bought it. In 1826, Märzel visited to US with Turk and opened an exhibition with the machine in New York.

The fact is, however, it was a human not the machine who played chess. The machine was tricky enough. A small human chess player sit inside the machine was not visible from outside when exhibitors showed inside the machine by opening the small door to make audiences confident this is a machine. The small human chess player hid himself by sliding his seat to the other small invisible place in the machine.

The secret had been perfectly kept for more than 100 years. It was not until Dr. Silas Mitchell fully revealed the secret in the book *'The Last of a Veteran Chess Player'* in 1857. No one had claimed that it was a human during these 100 years.

This episode might remind us the Turing Test. Besides his original imitation game, now called Turing Test, he also proposed its chess version, that is, Turing's chess machine. The game was with three chess players A, B and C. A operates the paper machine (since no computer existed at that time). A and B are in the separate room. C plays chess from outside the room with either A or B. Then C should guess whether he is playing, with human or the paper machine.

2   Turing Test

In mid February in 2011, IBM's room size supercomputer called Watson challenged 'Jeopardy' - America's favorite quiz show. In Jeopardy, normally three human contestants fight to answer questions over various topics, with a penalty for a wrong answer. The questions are like *"Who is the 19th-century painter whose name means police officer?"* or *"What is the city in US whose largest airport is named for a World War II hero; and its second largest for a World War II battle?"*

The contests were held over three days with Watson being one of the three contestants and the other two being the ex-champions of Jeopardy - Ken Jennings and Brad Rutter. As Watson cannot see or hear, questions were shown as a text file at the same moment when the questions were revealed to the two human contestants. By the end of the third days, Watson got \$77,147 while Jennings got \$24,000 and Rutter \$21,600. Watson beat the two human ex-champions.[1] If we set up an appropriate scenario, Watson could pass the Turing Test.

Or, in mid March in 2012, a computer program, called Dr. Fill, challenged 600 humans of the world's best crossword players at the American Crossword Puzzle Tournament in Brooklyn. Dr. Fill was created by Matthew Ginsberg aiming specifically to solve crossword puzzles. At the tournament, players will get six puzzles to solve on Saturday, and one on Sunday - progressively more difficult. Rankings are determined by accuracy and speed. The top three finishers enter a playoff with an eighth puzzles on Sunday afternoon, competing for the \$5,000 prize. The trophy went to a human and the computer program finished 141st.[2] Nevertheless, it was an impressive result, was it not?

In such an era when computer technology is so magnificent, it would not be difficult to imitate an human intelligence. One of the easiest ways to make a human believe that the machine is a human, might be a deliberate mistake from time to time pretending not to be very precise to imitate a human. Even at that time, Turing (1950) wrote,

It is claimed that the interrogator could distinguish the machine from the man simply by setting them a number of problems in arithmetic. The machine would be unmasked because of its deadly accuracy. The reply to this is simple. The machine (programmed for playing the game) would not attempt to give the right answers to the arithmetic problems. It would deliberately introduce mistakes in a manner calculated to confuse the interrogator.

Later, Michie (1993) wrote, *"Intelligence might be well demonstrated by concealing it, ... concerning this Turing's suggestion of machine's deliberate mistakes encouraged in order for the machine to pass the test."*

Turing Test is, to simply put, a test to know whether computer can fool human that "I am a human not a computer!" In recent years, we have a very practical program called *CAPTCHA* in order to prove "I'm not a computer but a human." Actually it stands for *'Completely Automated Public Turing Test to tell Computers and Humans Apart.'* This is an acronym based on the English word *'capture.'* This is sometimes called a reverse Turing Test.

A poker playing robot must cheat a web casino site to play there as human. Actually Hingston (2009) proposed a new test as follows:

Suppose you are playing an interactive video game with some entity. Could you tell, solely from the conduct of the game, whether the other

---

[1] This is from the article in New York Times by John Markoff entitled *"Creating artificial intelligence based on the real thing,"* on 17 February 2011.

[2] This is from the two articles in New York Times by Steve Lohr. One is *"The computer's next conquest: crosswords,"* on 17 March 2012, and the other is *"In crosswords, it's man over machine, for now,"* on 19 March 2012.

entity was a human player or a bot? If not, then the bot is deemed to have passed the test.

Despite those reflections above, it is strange to know that no computer algorithm has never succeeded in passing the Turing test. We have a contest organized by Hugh Loebner who will pledge $100,000 to the program that succeeds in passing the Turing Test if appeared.[3] The contest started in 1990. Four human judges sit at computer terminals with which the judges can talk both to the program and to the human who tries to mimic computer. Both are in another room and after, say, five minutes the judges must decide which is the person and which is the computer. The first computer program that judges cannot tell which is which will be given the award, and then this competition will end. As of today the contest has not ended yet.

Nowadays, it seems that the Turing Test is only of theoretical interest mainly from philosophical point of view. As a matter of fact, in computer science/ technology, what we are interested in might not be a binary decision like the Turing Test, that is, intelligent or not, but a degree to how intelligent it is.

We now proceed to more recent definitions of machine intelligence.

3   Formal Definitions of Machine Intelligence

This section starts with a survey of Legg and Hutter's elegant formal definition of machine intelligence (Legg and Hutter, 2007). It exploits Solomonoff's theory to predict a binary sequence of which prior distribution is unknown (Solomonoff, 1964), as well as Kolmogorov complexity theory. This trend, as Legg and Hutter wrote, was originated by Chaitin who used Gödel's complexity theory to define a machine intelligence. Chaitin wrote, *"(We hope to) develop formal definitions of intelligence and measures of its various components; apply information theory and complexity theory to AI,"* as directions for his future research (Chaitin, 1982). Legg and Hutter also pointed out Smith's proposal (Smith, 2006) as *"another complexity based formal definition of intelligence that appeared recently in an unpublished report."*

In order for the definition to be universal, Legg and Hutter creates infinitely large number of environments from one to the next, representing them by the universal Turing machine, in each of which an agent's intelligence is measured. This is not a very easy task. Hence, later, Hernández-Orallo (2010) suggested the other methods of representation for an environment describing that they are *"hopefully unbiased universal."* He also suggested a usage of finite state machine, admitting its representation is not Turing complete. A finite state machine was also used by Hibbard (2009). Hibbard also suggested a possible flaw of Legg and Hutters' way of testing an agent with *an infinitely large number of environments*, noting that it might be suffered from a result of the *'No Free Lunch theorem'* which claims *"any two algorithms are equivalent when their performance is averaged across all possible problems,"* (Wolpert and Macready 1997).

---

[3] http://www.loebner.net/Prizef/loebner−prize.html.

3.1   Legg and Hutter's Universal Intelligence of an Agent

Legg and Hutter had started an informal definition of intelligence for their formal definition to be based on. That is,

an ability to achieve goals in a wide range of environments.

So, an agent is given a series of wide range of environments, one by one, to be measured how intelligently the agent behaves seeking a goal in each of the environments given. The agent makes an *action* in the environment. The environment indicates options of how the agent can behave next, which is called *observation*, and reveals an information of how good or bad the agent's action is, which is called a *reward*. These actions, observations, and rewards are represented by symbols a, o, r from a finite set of A, O, and R, respectively. Thus, an agent interacts with its environment at each time in a sequence of discrete times, sending a_i to the environment and receiving o_i and r_i from the environment at time i. Repeating this procedure yields a sequence of action-observation-reward, such as:

$$a_1 o_1 r_1 a_2 o_2 r_2 a_3 o_3 r_3 \cdots, \tag{1}$$

called a *history*.

Now let's define an agent π. The agent is defined as a function that takes the current history as input and decides the next action as output. Under this definition the agent behavior is deterministic. If non-deterministic behavior is preferable, π can be a probability measure of the next action given the history before the action. Thus, for example,

$$\pi(a_3 | a_1 o_1 r_1 a_2 o_2 r_2) \tag{2}$$

is either a function or a probability measure of the 3rd action of the agent.

Similarly, environment μ is defined as a function or probability measure of $o_k r_k$ given the current history, that is,

$$\mu(o_k r_k | a_1 o_1 r_1 a_2 o_2 r_2 \cdots a_{k-1} o_{k-1} r_{k-1} a_k). \tag{3}$$

The formal measure of success of an agent π under the environment μ denoted as V_μ^π is defined as the expected value of the sum of rewards. That is,

$$V_\mu^\pi = E(\sum_{i=1}^{\infty} r_i). \tag{4}$$

This is the achievement of the goal in one environment.[4]

---

[4] Note here that r_i must be chosen such that V_μ^π ranges from 0 to 1. In the closely related concept of reinforcement learning, this is denoted as $E(\sum_{i=1}^{\infty}(1-\gamma)\gamma^{i-1}r_i)$ where $0 \leq \gamma \leq 1$ is called a *discount factor*, meaning "future rewards should be discounted," and $\sum_{i=1}^{\infty}(1-\gamma)\gamma^{i-1} = 1$. Later, Goertzel (2011) pointed out "discounting is so that near-term rewards are weighted much higher than long-term rewards," as being a not necessarily favorable outcome.

Then intelligence $\gamma(\pi)$ is defined as a weighted sum of this expected value of the sum of rewards over infinitely large number of various environments.

$$\gamma(\pi) = \sum_{\mu \in E} w_\mu \cdot V_\mu^\pi, \tag{5}$$

where E is the space of all environments under consideration.[5]

How will those weights be specified? Firstly, for the purpose, we describe the environment μ_i as a binary string x by a simple encoding algorithm via the universal Turing machine U. Second, we calculate Kolmogorov complexity of x as the length of the shortest program that computes x. This is the measure of the complexity of an environment. That is,

$$K(x) = \min_p \{l(p)|U(p) = x\}, \tag{6}$$

where p is a binary string which we call a program, $l(p)$ is the length of this string in bits, and U is a prefix universal Turing machine.

Thus the complexity of μ_i is expressed by $K(\mu_i)$. We use this in the form of probability distribution $2^{-K(\mu)}$. Then finally,

$$w_\mu = 2^{-K(\mu)}. \tag{7}$$

In case where we have a multiple paths to the goal, the simplest one should be preferred, which is sometimes called the principle of Occam's razor. In other words, *Given multiple hypotheses that represents the data, the simplest should be preferred,* as Legg and Hutter wrote. The above definition of the weight value follows this principle, that is, the smaller the complexity the larger the weight.

In summary, the intelligence of the agent π is the expected performance of agent with respect to the universal distribution $2^{-K(\mu)}$ over the space of all environments E, that is,

$$\gamma(\pi) = \sum_{\mu \in E} 2^{-K(\mu)} \cdot V_\mu^\pi. \tag{8}$$

In other words, weighted sum of the formal measure of success in all environments where the weight is determined by the Kolmogorov complexity of each environment.

We now recall the starting informal definition: *an ability to achieve goals in a wide range of environments.* In the above equation, *the agent's ability to achieve goals* is represented by V_μ^π, and *a wide range of environments* is represented by the summation over E, that is, all the well-defined environments in which rewards can be summed. Occam's razor principle is given by the factor $2^{-K(\mu)}$. Thus Legg and Hutter called this *the universal intelligence* of agent π.

---

[5] Legg and Hutter described this as "the summation should be over the space of all *computable reward-summable environments*," meaning the total amount of rewards the environment returns to any agent is bounded by 1.

Legg and Hutter concluded that *"Essentially, an agent's universal intelligence is a weighted sum of its performance over the space of all environments. Thus, we could randomly generate programs that describe environmental probability measures and then test the agent's performance against each of these environments. After sampling sufficiently many environments the agent's approximate universal intelligence would be computed by weighting its score in each environment according to the complexity of the environment as given by the length of its program. Finally, the formal definition places no limits on the internal workings of the agent. Thus, we can apply the definition to any system that is able to receive and generate information with view to achieving goals."*

3.2 Goertzel's Pragmatic Intelligence

To be a little more realistic than Legg and Hutter's definition, Goertzel (2011) extends the definition by adding (i) the distribution function ν that assigns each environment a probability; (ii) a goal function that maps finite sequence $a_1 o_1 r_1 a_2 o_2 r_2 \cdots a_i o_i$ into r_i; (iii) a conditional distribution γ, so that $\gamma(g, \mu)$ gives the weight of a goal g in the context of a particular environment μ where g is given by a symbol from G; (iv) the Boolean value $\tau_{g,\mu}(n)$ that tells whether it makes sense when we evaluate performance on goal g in environment μ over a period of n time steps with 1 meaning yes and 0 meaning no. If the agent who is provided with goal g in environment μ during a time-interval $T = \{t_1, \cdots, t_2\}$ then the expected goal-achievement of the agent during the time-interval is:

$$V_{\mu,g,T}^{\pi} = \sum_{i=t_1}^{t_2} r_i. \tag{9}$$

Thus, finally the intelligence of an agent π is defined as:

$$\Pi(\pi) = \sum_{\mu \in E, g \in G, T} \nu(\mu) \gamma(g, \mu) \tau_{g,\mu}(T) V_{\mu,g,T}^{\pi} \tag{10}$$

which Goertzel call a *pragmatic general intelligence*. Let me skip a detail here.

3.3 Hernándes-Orallo's Other Representations of Environment

Representing environments by Turing machine would be conceptual or theoretical, and it would be difficult, if not at all.

Hernándes-Orallo (2010) wrote, *"Apart from Turing machines, we have many other Turing complete models of computation, such as λ-calculus, combinatory logic, register machines, abstract state machines, Markov algorithms, term-rewriting systems, etc. Many of them are more natural and easy to work with than Turing machines."* In fact, Hernández-Orallo used a type of register machine to measure an agent's sequential inductive ability. See, e.g., (Hernández-Orallo 2000).

Thus, *"the environments can be generated and their complexity can be computed automatically,"* as Hernándes-Orallo wrote.

Hernándes-Orallo also describes a possibility to use a finite state machine, noting *"finite-state machines are not Turing-complete, though."*

3.4 Hibbard's Intelligence Definition of a Specific Agent

Hibbard (2009) proposed to measure intelligence of a specific agent, not an agent in general, who performed a task under a specific type of environment, not *a wide range of different environments* as Legg and Hutter's. Both agent and environment are represented by its own finite state machine as more realistic models than Turing machines.[6] He proposed to measure an agent in *"a hierarchy of sets of increasingly difficult environments."* Then an agent's intelligence is defined as *"the ordinal of the most difficult environment it can pass,"* as well as *"the number of time steps required to pass the test."*

Hibbard specifically set a scenario as prediction game between *predictor p* as an agent, and *evader e* as an environment, assuming action, observation, reward are all binary number from $B = \{0, 1\}$.

The finite state machine M_p for an predictor p has a state set S_p, an initial state I_p, and a mapping

$$M_p : B \times S_p \to S_p \times B, \tag{11}$$

meaning that predictor receives an input from the evader and state is transferred to a new state making an action.

Similarly, the finite state machine M_e for evader e, has state set S_e, initial state I_e, and mapping:

$$M_e : B \times S_e \to S_e \times B, \tag{12}$$

meaning that the evader see the action of the predictor as an input and transfer the current state to a new state and give a reward to the predictor.

Evader e creates a finite binary sequence $x_1 x_2 x_3 \cdots$, and predictor p creates also a finite binary sequence $y_1 y_2 y_3 \cdots$. A pair of evader e and predictor p interacts where e produces the sequence according to

$$x_{n+1} = e(y_1 y_2 y_3 \cdots y_n), \tag{13}$$

and p produces the sequence according to

$$y_{n+1} = p(x_1 x_2 x_3 \cdots x_n). \tag{14}$$

Then predictor p wins the round $(n + 1)$ if $y_{n+1} = x_{n+1}$ (implies evader catches predictor) and evader e wins if $y_{n+1} \neq x_{n+1}$ (implies evader fails to catch predictor).

---

[6] Hibbard also showed Turing machine version of this speculation in the same paper, but we only show its finite state machine version here.

The predictor p is said to learn to predict the evader e if there exists $k \in N$ such that $\forall n > k$, $y_n = x_n$ and the evader e is said to learn to evade the predictor p if there exists $k \in N$ such that $\forall n > k$, $y_n \neq x_n$.

The reward $r_n = 1$ when $y_n = x_n$ and $r_n = 0$ when $y_{n+1} \neq x_{n+1}$. Furthermore, we say the agent p passes at environment e if p learns to predict e.

Denote the number of states in S_e and S_p as $t(e)$ and $t(p)$, respectively. Also denote E_m as the set of evaders e such that $t(e) \leq m$, and P_m as the set of predictors p such that $t(p) \leq m$, given any $m \in N$.

Then Hibbard showed, with a proof, that there exists a predictor p_m that learns to predict all evaders in E_m, and there exists an evader e_m that learns to evade all predictors in P_m.

Finally, the intelligence of an agent p is measured as the greatest m such that p learns to predict all $e \in E_m$.[7]

4   Toward our Own Definition

Are we happy with above mentioned formal definitions of machine intelligence?

In a future, we might say, "This robot is universally more intelligent than that," when we have almighty humanoid robots, or androids, everywhere around us. Then universal intelligence will be interesting as we might expect super intelligent robot. At least at this moment, however, what we are interested in would be, *"How intelligently an agent completes a specific task?" "Is this agent more intelligent than that when the both challenge a specific task?"* or *"Which of the agents is more intelligent in doing this task?"*

4.1   Should Machine Intelligence be Universal?

Let's be specific not universal.

We could claim, *"She is an intelligent dancer,"* while we know she is not good at mathematics, or, *"He is an intelligent football player,"* while we know he is not good at physics, which we don't care. We can say, *"This conductor always makes an intelligent interpretation of symphony, but very bad at football."* Did Einstein play football intelligently?

The same holds. Machine intelligence doesn't need to be universal, at least at this moment. For instance, a cooking robot could be said to be very intelligent.

4.2   Should Intelligence be Ultra Efficient and Never be Erroneous?

Human intelligence is rather flexible than ultra-efficient.

Some of what they call intelligent machines may indeed perform the given task much more efficiently, effectively, or precisely than human do, while we human are not usually very efficient, effective nor precise, but rather spontaneous, flexible, unpredictable, or even erroneous sometime.

---

[7] Hibbard also took time step t within which agent p achieves intelligence n into consideration. But let me skip it here. See (Hibbard, 2009).

We don't necessarily expect artificial intelligence to be as efficient as human, but sometimes expect its flexibility, spontaneity, or unpredictability. Frosini (2009) wrote, *"Contradiction can be seen as a virtue rather than as a defect. Furthermore, the constant presence of inconsistencies in our thoughts leads us to the following natural question: 'Is contradiction accidental or is it the necessary companion of intelligence?' "*

4.3   Intelligence Tends to Avoid a Similar Behavior

When we address a human-like intelligence, we expect somewhat of a different behavior than the one as we behaved before, or at least not exactly the same one as before, even when we come across a same situation again.

Once my friend, who worked with a world famous electric company as an expert engineer, told me, *"It's amateurish,"* when I admired a food in a Chinese restaurant telling him, *"I think it's really wonderful that they cook every time in a slightly different way whenever I order the same one, and every time it's delicious."* He coldly told me, *"Real professional should cook exactly the same way every time."* Is this a condition for being intelligent?

Machine intelligence should not repeat same action even in a same situation as before!

Assume, for example, we are in a foreign country where we are not so conversant in its native language, and assume we ask, *"Pardon?"* to show we have failed to understand what they were telling us. Then intelligent people might try to change the expression with using easier words so that we understand this time, while others, perhaps not so intelligent, would repeat the same expression, probably a little louder.

A so-called smart robot-pet such as Sony's AIBO seems splendidly learns the environment of the owner. However, if it repeats an identical action in an identical situation, we would lose an interest sooner or later.

4.4   Should Intelligence Look Complex or Simple?

In his article in the TIME Magazine,[8] Jeffrey Kluger wrote, *"Another hypothesis is that it is actually intelligence that causes social relationships to become more complex, because intelligent individuals are more difficult to learn to know."* So, complexity might be one of the factors to measure intelligence.

At the same time, one of the factors to be intelligent is a capability to describe complex things clearly simple, that is, Occam's razor principle. As we have already seen, Legg and Hutter (2007) used Kolmogorov's complexity measure to incorporate this Occam's razor principle in their formal definition of machine intelligence.

4.5   Learning Capability

Learning is crucial for a human to be intelligent. So does a machine intelligence.

---

[8] The Time Magazine August 16, 2010, Vol. 176, No. 7 "Inside the minds of animals" (Science).

4.6   Toward our Own Formal Definition

First of all, let's restrict our definition simply to just one specific task. And let's forget measuring complexity by Turing machine or something like that, which is far from being pragmatic. Instead, let's look for some simpler complexity measure. Then, let's measure how an action is unpredictable, or spontaneous. A similarity of one action comparing to the previous actions should be also incorporated to the definition. Finally, as a measure of learning capability, by repeating the algorithm a multiple of times to observe how an action in a run has been improved from the one made in the previous runs.

Hence, the formula to know intelligence of an agent π for a task μ has a form like

$$V_\mu^\pi = \sum_{j=1}^{N} \sum_{i=1}^{M} F(a_{ij}) \cdot G(a_{ij}) \cdot H(a_{ij}) \cdot U(a_{ij}) \tag{15}$$

where a_{ij} is the i-th action in the j-th run. M is a total number of actions in a run, and N is a number of runs repeated in a same environment.

The function F represents complexity or simplicity, G is a measure of unpredictability, H is a similarity measure of an action comparing to the previous actions, and U measures how one action in a run is better or worth than the same situation in previous runs.

The function form of F is up to the philosophy of designer of this formula. The function G and H might be a slightly monotonic increasing function, the larger the better more or less. The function U is a monotonic decreasing function assuming we measure efficiency or time for an agent to process one task.

5   Conclusion

A couple of formal definitions of machine intelligence have been surveyed, after a reflection of what a real human-like intelligence should look like. Their approaches are elegant but still don't answer our question - "a degree to how a specific agent performs intelligently in a specific environment?" As such, yet another possible approach has been suggested from this point of view, though the idea is still not a matured one.

References

Chaitin, G.J.: Gödel's theorem and information. Theoretical Physics 21(12), 941–954 (1982)

Dreyfus, H.: What computers can't do. MIT Press (1979)

Frosini, P.: Does intelligence imply contradiction? Cognitive Systems Research 10(4), 297–315 (2009)

Gnilomedov, I., Nikolenko, S.: Agent-based economic modeling with finite state machines. In: Combined Proceedings of the International Symposium on Social Network Analysis and Norms for MAS, pp. 28–33 (2010)

Goertzel, G.: Toward a formal characterization of real-world general intelligence. In: Proceedings of the 3rd International Conference on Artificial General Intelligence, pp. 19–24 (2011)

Hernández-Orallo, J.: Beyond the Turing test. Journal of Logic, Language and Information 9(4), 447–466 (2000)

Hernández-Orallo, J.: A (hopefully) non-biased universal environment class for measuring intelligence of biological and artificial systems. In: Proceedings of the 3rd International Conference on Artificial General Intelligence, pp. 182–183 (2010)

Hernández-Orallo, J., Dowe, D.L.: Measuring universal intelligence: Towards an anytime intelligence test. Artificial Intelligence 174(18), 1508–1539 (2010)

Hibbard, B.: Bias and no free lunch in formal measures of intelligence. Journal of Artificial General Intelligence 1, 54–61 (2009)

Hibbard, B.: Measuring agent intelligence via hierarchies of environments. In: Schmidhuber, J., Thórisson, K.R., Looks, M. (eds.) AGI 2011. LNCS, vol. 6830, pp. 303–308. Springer, Heidelberg (2011)

Hingston, P.: A Turing test for computer game bots. IEEE Transactions on Computational Intelligence and AI in Games 1(3), 169–186 (2009)

Legg, S., Hutter, M.: Universal intelligence: A definition of machine intelligence. Minds and Machines 17(4), 391–444 (2007)

Malkiel, B.G.: A random walk down wall street: The time-tested strategy for successful investing. W.W. Norton & Company (2007)

McCorduck, P.: Machines who think: A personal inquiry into the history and prospects of artificial intelligence. A.K. Peters Ltd. (2004)

Michie, D.: Turing's test and conscious thought. Artificial Intelligence 60, 1–22 (1993)

Smith, W.D.: Mathematical definition of intelligence (and consequences) (2006), http://math.temple.edu/wds/homepage/works.html

Solomonoff, R.J.: A formal theory of inductive inference: Parts 1 and 2. Information and Control 7, 1–22, 224–254 (1964)

Standage, T.: The Turk: The life and times of the famous eighteenth-century chess-playing machine. Walker & Company (2002)

Turing, A.M.: Computing machinery and intelligence. Mind 59(236), 433–460, http://www.loebner.net/Prizef/TuringArticle.html

Turing, A.M.: Intelligent machinery (1948), See, e.g., Copeland, B.J.: The essential Turing: The ideas that gave birth to the computer age. Clarendon Press, Oxford (2004)

Weng, J.: Connectionists Digest 335(4), 12 (2013)

Wolpert, D., Macready, W.: No free lunch theorems for optimization. IEEE Transactions on Evolutionary Computation 1, 67–82 (1997)

Multi Objective Optimization of Trajectory Planning of Non-holonomic Mobile Robot in Dynamic Environment Using Enhanced GA by Fuzzy Motion Control and A*

Bashra Kadhim Oleiwi[1,*], Rami Al-Jarrah[1], Hubert Roth[1], and Bahaa I. Kazem[2]

[1] Siegen University/Automatic Control Engineering, Siegen, Germany,
Hoelderlinstr. 3
57068 Siegen
{bashra.kadhim,rami.al-jarrah,hubert.roth}@uni-siegen.de
[2] Mechatronics Eng. Dept University of Baghdad-Iraq
drbahaa@gmail.com

Abstract. a new hybrid approach based on Enhanced Genetic Algorithm by modified the search A* algorithm and fuzzy logic system is proposed to enhance the searching ability greatly of robot movement towards optimal solution state in static and dynamic environment. In this work, a global optimal path with avoiding obstacles is generated initially. Then, global optimal trajectory is fed to fuzzy motion controller to be regenerated into time based trajectory. When unknown obstacles come in the trajectory, fuzzy control will decrease the robot speed. The objective function for the proposed approach is for minimizing travelling distance, travelling time, smoothness and security, avoiding the static and dynamic obstacles in the robot workspace. The simulation results show that the proposed approach is able to achieve multi objective optimization in dynamic environment efficiently.

Keywords: Global and Local path planning, Trajectory generating, Mobile robot, Multi objective optimization, Dynamic environment, Genetic algorithm, Fuzzy control, A* search algorithm.

1   Introduction

The soft computing or intelligent systems include such as fuzzy logic, genetic algorithm, neural network can solve such complex within a reasonable accuracy [1, 2]. Motion planning [3] is one of the important tasks in intelligent control of an autonomous mobile robot. It is often decomposed into path planning and trajectory planning, although they are not independent of each other. Path planning is to generate a collision-free path in an environment with obstacles and to optimize it with respect to some criterion. Trajectory planning is to schedule the movement of a mobile robot along the planned path. There have been many methods proposed for motion planning of mobile robot [3]. Usually, motion planning of mobile robot under unknown environment was divided into two categories [4]. First, obstacles are unknown and

---

[*] Affiliated at University of Technology/Control and systems Eng.Dept. – Iraq.

V. Golovko and A. Imada (Eds.): ICNNAI 2014, CCIS 440, pp. 34–49, 2014.
© Springer International Publishing Switzerland 2014

their information of obstacles was completely unknown. It is the most important to avoid collision with obstacles for mobile robot, and not search the optimum motion planning. Second, the obstacles known and their information were known.

Many researchers have proposed various techniques for autonomous mobile robot navigation and obstacle avoidance to improve the robot performance. These approaches used the artificial intelligent algorithms such as fuzzy logic, neural networks and genetic algorithms [5-15]. During the past several years hybrid genetic-fuzzy approach has considered as one of the most powerful technique for research of intelligent system design application [12-15]. The Proposed controller in [12] was designed using fuzzy logics theory and then, a genetic algorithm was applied to optimize the scaling factors of the fuzzy logic controller for better accuracy and smoothness of the robot trajectory. In [13], they provided a hybrid genetic-fuzzy approach by which an improved set of rules governing the actions and behavior of a simple navigating and obstacle avoiding autonomous mobile robot.

The novel genetic fuzzy algorithm has applied in [14] to generate a dynamic path tracking for mobile robot in the unknown environment. Hence, genetic algorithm is used to find the optimal path for a mobile robot to move in a dynamic environment expressed by a map with nodes and links. a genetic-neuro-fuzzy strategy has been proposed in [15] to drive a mobile robot. This approach is able to extract automatically the fuzzy rules and the membership functions in order to guide a wheeled mobile robot. However, most of these researches solved the mobile robot navigation and obstacles avoidance problem with complicated mathematical computations and the planned optimal path was single objective. Even though few researchers studied the multi objective problem, but they ignore some important aspects such as minimum travelling time for the trajectory generating, safety factor, and security as a multi object optimization.

In this paper, we extended our approach in [16, 17], Here we address the multi objective optimization problem (MOOP) of mobile robot navigation and obstacles avoidance. The formulation of the problem is concerned with finding three important steps. The first step is to find the multi objectives optimal path for mobile robot from its start and target position with collision free of static obstacles. Hence, the multi objectives function is concerned with minimizing the traveling distance, time, and maintaining the smoothness and safety requirements of path by using enhanced genetic algorithm by modified A* and classical method in initialization stage and meanwhile, several genetic operators have been proposed according to characteristics of path planning and domain-specific knowledge of problem to overcome GA drawbacks such as the initial population in the traditional GA is generated randomly, which it leads to the huge size of population, a large search space, poor ability to rid of the redundant individuality, and the speed and accuracy of path planning is not satisfactory, especially in the complex environment and multi objectives planning. In addition, when the environment is complex and the number of the obstacles is increasing, the basic GA may face some difficulties to find a solution or even it may not find one. In order to deal with this issue, the second step was used to convert optimal path to trajectory. This means a time-based profile trajectory of position and velocity from start to destination. To maximize the distance between a mobile robot and its nearest dynamic obstacle, the fuzzy motion planning includes the fuzzy motion controller which helps the mobile robot to avoid dynamic obstacle was presented. These three steps are put together to build a complete unity and objective optimization solution of mobile robot navigation and obstacles avoidance.

The article is organized as follow: section 2 describes the problem and case study. Section 3 is devoted to Kinematics model of mobile robot and then the proposed approach, flow charts and evaluation criteria are introduced in section 4. The Multi objectives optimization function is addressed in section 5. The fuzzy motion planning was presented in section 6. Based on this formulation, section 7 is presented simulation results. Finally, in section 8 conclusions and future work are discussed.

2   Problem Description

As it is known, the path planning is used to generate a path off-line from initial to final position. On other hand, the trajectory generation is to impose a velocity profile to convert the path to a trajectory. A trajectory is a path which is an explicit function of time. Fig. 1 shows the motion planning of mobile robot. Finding a feasible trajectory is called trajectory planning or motion planning [18].

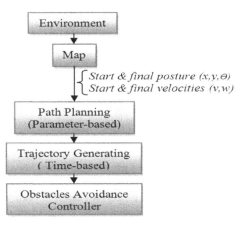

Fig. 1. Trajectory generation of mobile robot

Note that the trajectory means a time-based profile of position and velocity from start to destination while paths are based on non-time parameters. It is used to have smooth movement during the parametric Cubic Spline function of trajectory, and it must give continuous velocity and acceleration. Basically path planning problem is a form of geometrical problem which can be solved by geometrical description of mobile robot and its workspace, starting and target configuration of mobile robot and evaluation of degrees of freedom (DOF) of mobile robot [19]. Trajectory planning in general requires a path planner to compute geometric collision free path. This free path is to be converted into a time based trajectory [19]. The Kinematics model of the used mobile robot is presented as shown in the following section. Given a point mobile robot moves in 2D static and dynamic environment. These obstacles can be placed at any grid point in the map. The mobile robot's mission is to search offline and online optimal path that travels from a start point to a goal point starting point and the goal point in both environments that complies with the some restrictions. First, collision-free which means there should be no collision with the static and dynamic obstacles that appear on its way. Then, the short that minimize traveling distance. Third, smooth in order to minimize total angles of all vectorial path segments and minimum curvature. Fourth, security (safe, clear) to maintain the clearance requirements, the path is safer and farthest from obstacles (it should not approach to the obstacles very closely) or maximum clearance distance from obstacles should be keep to. The shortest time to minimize time traveled. Then, to minimize energy consumption of robot. Finally, the path must stay inside the grid boundaries.

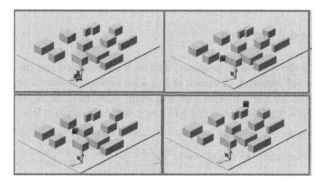

Fig. 2. Problem Description

As it is shown in Fig.2, when the robot navigates from start point to the target point, the multi objectives optimal trajectory planning which has been presented in this work is trade off among all of the aforementioned objectives with the advantage that multiple tradeoff solutions can be obtained in a single run.

3 Kinematic Model of the Robot

It assumes that the robot is placed on a plane surface. The contacts between the wheels of the robot and the rigid horizontal plane have pure rolling and non slipping conditions during the motion. A simple structure of differentially driven three wheel robot as shown in Fig. 3. More details about Kinematics model and nonholonomic constraint can be found in [20, 21]. For the trajectory function parametric Cubic Spline function has been used [8]. For the boundary conditions, the wheeled mobile robot has to start its motion from its initial position, attains a suitable speed and must reach its goal position.

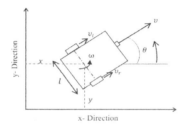

Fig. 3. Posture Coordinates of Mobile Robot [20]

The boundary conditions including the position, velocity and acceleration constraints of robot. the position constraint of the mobile robot[8].

$$x(0) = x_{initial} \text{ and } x(t_f) = x_{target} \qquad (1)$$

Velocity and acceleration constraints are that the mobile robot should start from rest at initial position with certain acceleration to reach its maximum velocity and near the target location it should decelerate to stop at the goal position.

4   Proposed Approach

In this section, a description of proposed approach is presented and the general schema of the proposed approach is shown in Fig 4 as well as the flow chart in Fig. 5. In order to solve the MOOP problem can five main steps have been introduced. Note that the fifth step is the fuzzy motion logic which will be described later.

Fig. 4. General scheme of the methodology [22]

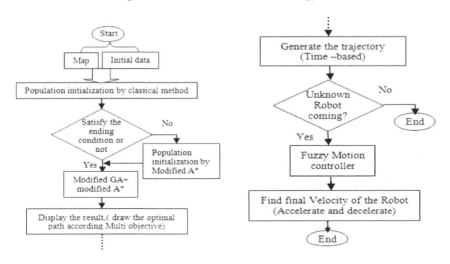

Fig. 5. Flowchart of proposed approach

The first step is the initialization. There are some of definitions corresponding to the initialization stage are presented in Table 1. The next step describes the environment model and some corresponding definitions are presented. We construct a closed workspace (indoor area) with different numbers of static obstacles. This area is described by a 2D static map (20 × 20); the starting point is S=(1, 1), and the target point is T=(19, 19) for a path. The positions of the obstacles are randomly chosen; in

Table 1. Proposed approach parameter Specifications

GA Type	Modified GA
Population size	10
Chromosome length	Varies
Crossover type	One point crossover
Mutation type	Flip bits
Crossover rate, Pc	0.8
Mutation rate, Pm	0.35
Max. Iteration (i)	50

other words, the obstacles can be placed at any grid point in the map. In order to eliminate obstacles nodes from the map at the beginning of the algorithm a shortcut or decreased operator had been used.

In the third stage, which is called an initial population which generate and moving for sub optimal feasible paths. The classical method and modified A* is used for generating a set of the sub optimal feasible paths in both simple and complex maps. Then, the obtained paths are used for establishing the initial population for the GA optimization. Here, the mobile robot moves in an indoor area and it can move in any of the eight directions (forward, backward, right, left, right-up, right-down, left-up, and left-down). In the classical method, the movement of the mobile robot is controlled by a transition rule function, which in turn depends on the Euclidean distance between two points (the next position j and the target position T) and roulette wheel method to select the next point and to avoid falling in local min in complex map. Hence, the distance value (D) between two points is:

$$D = \begin{cases} \sqrt{2} & \text{if the robot moved its diagonal direction} \\ 1 & \text{otherwise} \end{cases} \tag{2}$$

The robot moves through every feasible solution to find the optimal solution in favored tracks that have a relatively less distance between two points, where the location of the mobile robot and the quality of the solution are maintained such that the sub optimal solution can be obtained. However, when the number of the obstacles is increasing, the classical method may face difficulties to find a solution or even they may not find one. Also, the more via points are used the more time consuming in the algorithm that depends mostly on the number of via points that they will use in the path in maze map. In the case of adding the modified A* search algorithm in initialization stage of GA, the proposed approach will find a solution in any case, even if there are many obstacles. In fact, the traditional A* algorithm is the standard search algorithm for the shortest path problem in a graph. The A* algorithm as shown in equation (3) below can be considered as the best first search algorithm that combines the advantages of uniform-cost and greedy searches using a fitness function [23].

$$F(n) = (g(n) + h(n)). \tag{3}$$

Where g(n) denotes the accumulated cost from the start node to node n and h(n) is a heuristic estimation of the remaining cost to get from node n to the goal node [23] . In our study the accumulated cost and heuristic cost are the Euclidean distance between two nodes. The robot selects the next node depends on the minimum value of F(n). The modification of A* algorithm is the most effective free space searching algorithms in term of path length optimization (for single objective). We proposed modified A* for searching of sub optimal feasible path regardless of length to establish the initial solution of GA in maze map, by adding the probability function to A* method. We have modified the A* in order to avoid use the shortest path which it could affect the path performance in term of multi objective (length, security and smoothness) in initial stage.

$$F(n) = Rand*(\ g(n) + h(n)).\qquad(4)$$

The next step is the optimization, more precisely it is modified GA. This step uses the modified GA for optimizing the search of the sub optimal path that generated in step 3. Hence, the main stages in modified GA are natural selection, standard crossover, proposed deletion operator , enhanced mutation with basic A* and sort operator to improve the algorithm's efficiency according to characteristics of path planning, because it is difficult to achieve convergence and generate optimized feasible paths only by operators in standard GA. The initial population of the potential solutions of the problem will be created by classical method or modified A* as the initial population, and it is called chromosomes. In the proposed approach, a chromosome represents the path and its length varies depending on the case at hand. This means that it consists of a set of genes (via-points) of the path from the start position to the target position. Since, $p(x_0,y_0) = (1, 1)$ is always the starting point and $p(x_n,y_n) = (19, 19)$ is always the target point, the via-points of the path are $p(x_1,y_1)$ and $p(x_{i+1},y_{i+1})$, and all these points are represented the genes of the chromosome as shown in Fig. 6.

$Chromosome\ or\ Path\ (P\ ) = \{(x_0, y_0),(x_1, y_1),\ldots\ldots,(x_{i+1}, y_{i+1}),\ldots\ldots,(x_n, y_n)\}$

Fig. 6. Chromosome structure

In each generation, all chromosomes will be evaluated by fitness function F which will be discussed later. Thus, a chromosome with the minimum fitness has a considerably higher probability than others to select and reproduce by means of GA operators in the next generation. These steps are repeated until the maximum number of iterations is reached as shown in flowchart Fig. 5.

-GA operators
- *Selection*: Two parents randomly are selected based on their fitness by using the Roulette wheel selection method
- *-Crossover Operator:* During the crossover operation, two chromosomes that have an efficient fitness values are randomly selected as parents based on the selection method. Hence, each parent has its chromosome length. In this study, single point crossover is used. Crossover operation swaps the two parents around crossover points: This operation results feasible paths, because the nodes before crossover point in the first parent and the nodes after crossover point in the second parent and in opposite, are valid nodes, as shown in Fig. 7a.

- *Mutation Operator:* The parental chromosome is chosen according to selection method. The parents start and target nodes are not mutated. The mutation operation is done by selecting an intermediate node in the parent according to mutation probability. These nodes are chosen randomly to replace the mutated node.
- *Enhanced Mutation operator*: It is served as a key role to diversity the solution population, we proposed to enhance mutation operator by adding traditional A* search method to mutation. The enhanced mutation method is used to avoid fall into a local minimum, improve and decrease the distance of the partially path, between two randomly points (i and j) included in the main path, as shown in Fig. 7b
- *Deletion operator:* It proposed to eliminate the repeated genes (redundant) from an individual (path). For specific gene, the approach reversely check if this is equal to others and this is done for each gene, as shown in Fig. 7c.
- *Sort operation:* This operator sorts the chromosomes of population according to their fitness at each generation. The feasible chromosomes are organized in ascending order according to their fitness, and secondly, if a group of chromosomes has an equal fitness values, they are again sorted, in ascending order.

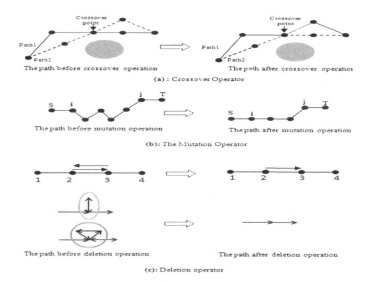

Fig. 7. GA Operators

5   Multi Objective Fitness Function

Real-world problem solving will commonly involve the optimization of two or more objectives at once. The consequence of this strategy is not always possible to reach an optimal solution with respect to all of the objectives evaluated individually. Historically a common method used to solve multi objective problems is by a linear combination of the objectives, in this way creating a single objective function to optimize or by converting the objectives into restrictions imposed on the optimization problem. In regards to evolutionary computation [24], proposed the first implementation for a multi objective evolutionary search. Most of proposed methods focus around the concept of Pareto optimality and the Pareto optimal set. Using these concepts of optimality of individuals evaluated under a multi objective problem, they propose a fitness assignment to each individual in a current population during an evolutionary search based upon the concepts of dominance and non-dominance of Pareto optimality [24, 25]. In recent years, the idea of Pareto-optimality is introduced to solve multi-objective optimization problem with the advantage that multiple tradeoff solutions can be obtained in a single run [26]. The total cost of fitness (or objective) function of feasible path P with n points is obtained by a linear combination of the weighted sum of multi objectives as follows [27, 28]:

$$\min F(P) = \min \{\omega_1 F_1(P) + \omega_2 F_2(P) + \omega_3 F_3(P) + \omega_4 F_4(P)\} \tag{5}$$

Where ω_1, ω_2, ω_3 and ω_4 represent the weight of each objective to total cost or multi objective function F(P). The remaining parameters are defined as follow [16, 17], Where, $F_1(P)$ is the total length of path and criteria of path, $F_2(P)$ is the path smoothness, $F_3(P)$ is the path clearance or path security. $F_4(P)$ represents the total consumed time for robot motion and it can obtained by:

$$F_4(P) = t_T \tag{6}$$

Where t_T is the total time from start to target point. By minimizing the overall fitness function regarding the assigned weights of each criterion, a suitable path is obtained. The weights of the shortest, smoothest and security fitness functions, w_1, w_2, w_3 and w_4 respectively, are tuned through simulation and try and errors, with best found values.

6   Fuzzy Motion Planning

The fuzzy logic controller is one of the most successful applications of the fuzzy set theory [29]. The performance of this controller depends on the selection of membership functions and rule base. Therefore, fuzzy logic system mainly includes fuzzification, rule base and fuzzy reasoning and defuzzification. In this work. the proposed fuzzy control contains an obstacle avoidance strategy and it is the final step in our approach. It has two inputs. The first input is the optimum velocity from the GA and the second one is the obstacle when the robot detects new moving obstacle in its path. The output is the final velocity of the robot. The fuzzy control might control the robot speed based on detection of any dynamic object. Not only the robot it can reduce the speed in case the dynamic object gets more close to robot, but also the control has ability to increase the speed when the robot becomes safe away from the dynamic obstacle. Fig 8 shows the navigation expression problem in the structure of fuzzy logic.

Fig. 8. Fuzzy logic algorithm

The membership functions for the proposed algorithms are shown in Fig 9, Fig 10 and Fig 11. According to Fig. 9 and Fig 10, fuzzy membership functions for input velocity has 9 linguistic variables. These linguistic variables are (Z: Zero, VVL: very very low, VL: very low, L: low, Medium, H: High, VH: very high, VVH: very very high, Maximum). The second input has 5 linguistic variables (No, Far, Medium, Close, Very Close). It should be noted that the input 2 is normalized. For the output 9 membership functions have been used (Z: Zero, VVL: very very low, VL: very low, L: low, Medium, H: High, VH: very high, VVH: very very high, Maximum).

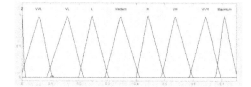
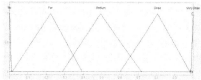

Fig. 9. Membership functions for the input Velocity

Fig. 10. Membership functions for the input detect obstacle

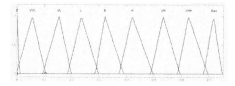

Fig. 11. Membership functions for the output final velocity

The most common defuzzification methods are including centroid and the weighted average methods. This step is an operation to produce a non-fuzzy control action. It transforms fuzzy sets into crisp value. Therefore, in this work, for the ultimate defuzzification the gravity method has been used as it is given by:

$$Z_a = \frac{\int \mu_c(z) \cdot z\, dz}{\int \mu_c(z)\, dz} \tag{7}$$

Where $\mu_c(z)$ is the degree of membership, z_a is the crisp value.

7   Simulation Results

This section presents the results of robot case study. In order to verify the effectiveness of the proposed hybrid approach, we applied it in simple and complicated 2D static environments with different numbers of obstacles. The MATLAB software

(CPU is 2.61 GHz) is used for the simulation. Fig. 13, Fig. 17 and Fig. 21 show the velocity profile for the optimal trajectory generation for different maps. The robot paths have been successful designed by GA in this cases there are no unknown obstacles appears in the paths. The optimal trajectory generation for these paths are shown in Fig 12, Fig 16 and Fig 20.the fuzzy velocity profile for the optimal Trajectory generation are shown in Fig 14, Fig. 18 and Fig. 22. The fuzzy motion controller have given approximately the similar GA velocity values for the robot. In these cases, we still have no unknown obstacles come to the robot path. Therefore, the fuzzy control and the GA work simultaneously in order to make the robot navigate to its target without colliding with obstacles. Fig. 15, Fig. 19 and Fig. 23 are show how this algorithm can reduce the velocity of the robot not only if there are static obstacles, but also in case there are moving objects come to the path. Fig. 15 shows the velocity of the robot with respect to time for optimum and it starts increasing from 0 to maximum 0.75 m/s and the robot might reduce its speed in case there is a dynamic obstacle close to its path in order to turn left or right. Fig. 19 shows the optimum robot velocity with respect to time. It starts accelerate from 0 to maximum 0.75 m/s and the robot has the ability to decrease its speed if a dynamic obstacle come close to its path as well as accelerate again back to its maximum velocity. In Fig. 23 shows the optimum velocity of the robot compares to the output velocity gained from the fuzzy control. The fuzzy logic not only could control this velocity but also might work if the robot detects a moving obstacle in its path. It is obvious clear that when the robot detects the obstacle at time 30 the fuzzy control will decrease the velocity to 0.55m/sec and increase it again to 0.75m/sec when the path gets free from the obstacle.

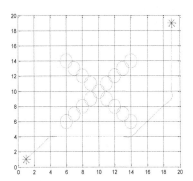
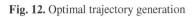

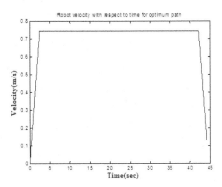

Fig. 12. Optimal trajectory generation

Fig. 13. Final Velocity Profile for Optimal Trajectory generation

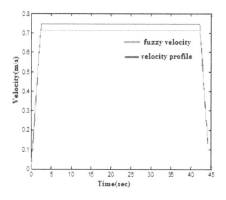

Fig. 14. Final Fuzzy Velocity Profile for the Optimal Trajectory generation

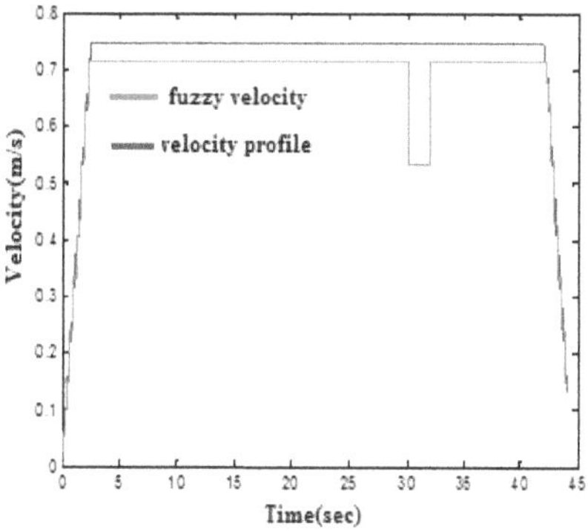

Fig. 15. The Fuzzy Velocity Profile with obstacle appears

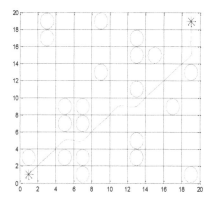

Fig. 16. Optimal trajectory generation

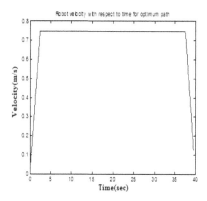

Fig. 17. The Final Velocity Profile for the Optimal Trajectory generation

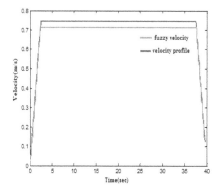

Fig. 18. Final Fuzzy Velocity Profile for the Optimal Trajectory generation

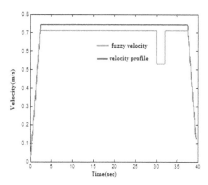

Fig. 19. The Fuzzy Velocity Profile with obstacle appears

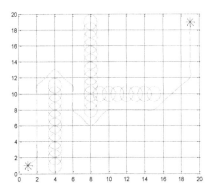

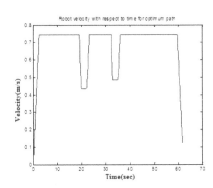

Fig. 20. Optimal trajectory generation

Fig. 21. The Final Velocity Profile for the Optimal Trajectory generation

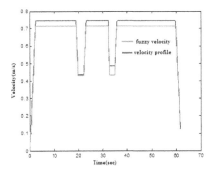

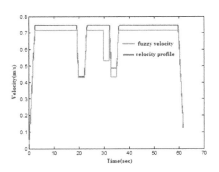

Fig. 22. Final Fuzzy Velocity Profile for the Optimal Trajectory generation

Fig. 23. The Fuzzy Velocity Profile with obstacle appears

As it is shown in Fig. 24 in order to avoid unknown moving object (red) in the environment the fuzzy motion controller will reduce the final velocity for the robot (green) when it detects new unknown obstacle. Hence the robot can avoid collision with the moving object and when the path becomes free again the controller will increase the velocity and the robot can finish its mission to reach the target.

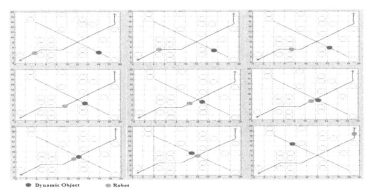

Fig. 24. Path planning with fuzzy logic in relatively complex map

8   Conclusion

The main contribution of this study is the presentation of a proposed approach to generate multi objective optimization of path and trajectory of mobile robot with free-collision. This approach is considering cubic spline data interpolation and the non-holonomic constrains in Kinematic equations of mobile robot, which is used to obtain a smooth trajectory with an associated minimum energy cost. In this work, a global optimal path with avoiding obstacles is generated initially. Then, global optimal trajectory is fed to fuzzy motion controller to be regenerated into time based trajectory. The fuzzy control shows a good performance to deal with dynamic obstacle in the environment. The objective function for the proposed approach is for minimizing travelling distance, travelling time, smoothness and security, avoiding the static and dynamic obstacles in the workspace. The simulation results show that the proposed approach is able to achieve multi objective optimization in dynamic environment efficiently.

Acknowledgment. The first author gratefully acknowledges German academic exchange service (DAAD), University of Siegen and, Ministry of Higher Education and Scientific Research in Iraq for financial support.

References

[1] Hui, N.B., Mahendar, V., Pratihar, D.K.: Time Optimal, Collision-Free Navigation of a Car-Like Mobile Robot Using Neuro-Fuzzy Approaches. Fuzzy Sets and Systems 157, 2171–2204 (2006)

[2] Jelena, G., Nigel, S.: Neuro-Fuzzy Control of a Mobile Robot. Neuro Computing 28, 127–143 (2009)

[3] Sugihara, K., Smith, J.: Genetic Algorithms for Adaptive Motion Planning of an Autonomous Mobile Robot. In: IEEE International Symposium on Computational Intelligence in Robotics and Automation, pp. 138–143 (1997)

[4] Lei, L., Wang, H., Wu, Q.: Improved Genetic Algorithms Based Path Planning of Mobile Robot Under Dynamic Unknown Environment. In: IEEE International Conference on Mechatronics and Automation, pp. 1728–1732 (2006)

[5] Li, X., Choi, B.-J.: Design of Obstacle Avoidance System for Mobile Robot Using Fuzzy Logic Systems. International Journal of Smart Home 7(3), 321–328 (2013)

[6] Li, X., Choi, B.-J.: Obstacle Avoidance of Mobile Robot by Fuzzy Logic System. In: ISA, ASTL, vol. 21, pp. 244–246. SERSC (2013)

[7] Purian, F.K., Sadeghian, E.: Path Planning of Mobile robots Via Fuzzy Logic in Unknown Dynamic Environments with Different Complexities. Journal of Basic and Applied Scientific Research 3(2s), 528–535 (2013)

[8] Arshad, M., Choudhry, M.A.: Trajectory Planning of Mobile robot in Unstructured Environment for Multiple Objects. Mehran University Research Journal of Engineering & Technology 31(1), 39–50 (2012)

[9] Rusu, C.G., Birou, I.T., Szöke, E.: Fuzzy Based Obstacle Avoidance System for Autonomous Mobile Robot. In: IEEE International Conference on Automation Quality and Testing Robotics, vol. (1), pp. 1–6 (2010)

[10] Rusu, C.G., Birou, I.T.: Obstacle Avoidance Fuzzy System for Mobile Robot with IR Sensors. In: 10th International Conference on Development and Application, pp. 25–29 (2010)

[11] Shi, P., Cui, Y.: Dynamic Path Planning for Mobile Robot Based on Genetic Algorithm in Unknown Environment: In: IEEE Conference on, pp. 4325–4329 (2010)

[12] Benbouabdallah, K., Qi-dan, Z.: Genetic Fuzzy Logic Control Technique for a Mobile Robot Tracking a Moving Target. International Journal of Computer Science Issues 10(1), 607–613 (2013)

[13] Senthilkumar, K.S., Bharadwaj, K.K.: Hybrid Genetic-Fuzzy Approach to Autonomous Mobile Robot. In: IEEE International Conference on Technologies for Practical Robot Applications, pp. 29–34 (2009)

[14] Farshchi, S.M.R., NezhadHoseini, S.A., Mohammadi, F.: A Novel Implementation of G-Fuzzy Logic Controller Algorithm on Mobile Robot Motion Planning Problem. Canadian Center of Science and Education, Computer and Information Science 4(2), 102–114 (2011)

[15] Phinni, M.J., Sudheer, A.P., RamaKrishna, M., Jemshid, K.K.: Obstacle Avoidance of a wheeled mobile robot: A Genetic-neurofuzzy approach. In: IISc Centenary – International Confonference on Advances in Mechanical Engineering (2008)

[16] Oleiwi, B.K., Hubert, R., Kazem, B.: Modified Genetic Algorithm based on A* algorithm of Multi objective optimization for Path Planning. In: 6th International Conference on Computer and Automation Engineering, vol. 2(4), pp. 357–362 (2014)

[17] Oleiwi, B.K., Roth, H., Kazem, B.: A Hybrid Approach based on ACO and GA for Multi Objective Mobile Robot Path Planning. Applied Mechanics and Materials 527, 203–212 (2014)

[18] Kim, C.H., Kim, B.K.: Minimum-Energy Motion Planning for Differential-Driven Wheeld Mobile Robots. Motion Planning Source in Tech. (2008)

[19] Breyak, M., Petrovic, I.: Time Optimal Trajectory Planning Along Predefined Path for Mobile Robots with Velocity and Acceleration Constraints. In: IEEE/ASME International Conference on Advanced Intelligent Mechatronics, pp. 942–947 (2011)

[20] Vivekananthan, R., Karunamoorthy, L.: A Time Optimal Path Planning for Trajectory Tracking of Wheeled Mobile Robots. Journal of Automation, Mobile Robotics & Intelligent Systems 5(2), 35–41 (2011)

[21] Xianhua, J., Motai, Y., Zhu, X.: Predictive fuzzy control for a mobile robot with nonholonomic constraints. In: IEEE Mid-Summer Workshop on Soft Computing in Industrial Applications, Helsinki University of Technology, Espoo, Finland (2005)

[22] Buniyamin, N., Sariff, N.B.: Comparative Study of Genetic Algorithm and Ant Colony Optimization Algorithm Performances for Robot Path Planning in Global Static Environments of Different Complexities. In: IEEE International Symposium on Computational Intelligence in Robotics and Automation, pp. 132–137 (2009)

[23] Buniyamin, N., Sariff, N.B.: An Overview of Autonomous Mobile Robot Path Planning Algorithms. In: 4th IEEE Student Conference on Research and Development, pp. 183–188 (2006)

[24] Krishnan, P.S., Paw, J.K.S., Tiong, S.K.: Cognitive Map Approach for Mobility Path Optimization using Multiple Objectives Genetic Algorithm. In: 4th IEEE International Conference on Autonomous Robots and Agents, pp. 267–272 (2009)

[25] Castillo, O., Trujillo, L.: Multiple Objective Optimization Genetic Algorithms for Path Planning in Autonomous Mobile Robots. International Journal of Computers, Systems and Signals 6(1), 48–63 (2005)

[26] Fonseca, C.M., Fleming, P.J.: An Overview of Evolutionary Algorithms in Multi-objective Optimization. Evolutionary Computing 3(1), 1–16 (1995)

[27] Jun, H., Qingbao, Q.: Multi-Objective Mobile Robot Path Planning based on Improved Genetic Algorithm. In: Proc. IEEE International Conference on Intelligent Computation Technology and Automation, vol. 2, pp. 752–756 (2010)

[28] Geetha, S., Chitra, G.M., Jayalakshmi, V.: Multi Objective Mobile Robot Path Planning based on Hybrid Algorithm. In: 3rd IEEE International Conference on Electronics Computer Technology, vol. 6, pp. 251–255 (2011)

[29] Saffiotti, A.: The use of fuzzy logic for autonomous robot navigation. Soft Computing 1(4), 180–197 (1997)

Multi Objective Optimization of Path and Trajectory Planning for Non-holonomic Mobile Robot Using Enhanced Genetic Algorithm

Bashra Kadhim Oleiwi[1,*], Hubert Roth[1], and Bahaa I. Kazem[2]

[1] Siegen University/Automatic Control Engineering, Siegen, Germany,
Hoelderlinstr. 3
57068 Siegen
{bashra.kadhim,hubert.roth}@uni-siegen.de, drbahaa@gmail.com
[2] Mechatronics Eng. Dept University of Baghdad-Iraq

Abstract. a new hybrid approach based on a modified Genetic Algorithm (GA) and a modified search algorithm (A*) is proposed to enhance the searching ability of mobile robot movement towards optimal solution state in static environment, and to achieve a multi objectives optimization problem of path and trajectory generating. According to that the cubic spline data interpolation and the non-holonomic constrains in Kinematic equations for mobile robot are used. The objective function of the proposed approach is to minimize traveling distance, and traveling time, to increase smoothness, security, and to avoid collision with any obstacle in the robot workspace. The simulation results show that the proposed approach is able to achieve multi objective optimization efficiently in a complex static environment. Also, it has the ability to find a solution when the number of obstacles is increasing. The mobile robot successfully travels from the starting position to the desired goal with an optimal trajectory as a result of the approach presented in this paper.

Keywords: path planning, trajectory generating, mobile robot, Multi objective optimization, static complex environment, GA, A*.

1   Introduction

Real-world problem solving commonly involve the optimization of two or more objectives at once. A consequence of this is that it is not always possible to reach an optimal solution with respect to all of the objectives evaluated individually. Historically a common method used to solve multi objective problems is a linear combination of the objectives, creating a single objective function to optimize, or converting the objectives into restrictions imposed on the optimization problem. With regards to evolutionary computation [1], the first implementation of a multi objective evolutionary search is proposed. Most of the proposed methods focus on the concept of Pareto optimality and the Pareto optimal set. Using these concepts of optimality of

---

* Affiliated at University of Technology/Control and systems Eng.Dept. – Iraq.

V. Golovko and A. Imada (Eds.): ICNNAI 2014, CCIS 440, pp. 50–62, 2014.
© Springer International Publishing Switzerland 2014

individuals evaluated under a multi objective problem, they each propose a fitness assignment for each individual in a current population during an evolutionary search based upon the concepts of dominance and non-dominance of Pareto optimality [1].

At present, mobile robotic path planning method uses two types of methods: traditional methods and intelligent methods. The GA is an excellent and mature intelligent method and is popular with the majority of researchers. The initial population in the traditional GA is generated randomly. However, the huge population size leads to a large search space, poor removal of redundant individuality, and unsatisfactory speed and accuracy of path planning, especially in complex environment, multi-robot path planning [2] and multi objectives. When the environment is complex and the number of the obstacles is increasing, the basic GA may have difficulties finding a solution or even may not find one at all. In addition to that it is difficult to achieve convergence and generate optimized and feasible multi objective path planning by using only the basic operators in standard GA [3]. A multi-objective mobile robot path planning is a wide and active research area, where many methods have been applied to tackle this problem [4-9]. A multi-objective GA robot trajectory planner is proposed in [4]. The hybrid gravitational search algorithm and a particle swarm optimization algorithm in [5] solved the optimization problems by minimizing the objective functions, producing optimal collision-free trajectories in terms of minimizing the length of the path that needs to be followed by the robot and also assuring that the generated trajectories are at a safe distance from the danger zones. In [6] the described the use of a GA for the problem of offline point-to-point mobile robot path planning. The problem consist of generating "valid" paths or trajectories, represented by a two dimensional grid, with obstacles and dangerous ground that the robot must evade. This means that the GA optimizes possible paths based on two criteria: length and difficulty. Actually, some of these researches solved the multi objective optimization for mobile robot path planning problem without taking into account some important issues such as minimum travelling time for the trajectory generating [7, 8]. Even though a few researchers studied the multi objective problem included minimum travelling time, but they ignore some important aspects such safety factor and security as a multi object optimization for the trajectory generating.

In this work, we extend our approach in [10] by taking into account the travelling time as a fourth objective as well as reducing the energy consumption of mobile robot. The approach based on performing modification of A* and GA to enhance the searching ability of robot movement to reach an optimal solution in term of multi objective optimal path and trajectory for mobile robot navigation in complex static environment. In addition, the traditional GA has some drawbacks such as huge population size, large search space, and poor ability of removing redundant individuality. Also, when the environment is complex and the number of the obstacles is increasing, the GA might not find a solution or it can face some difficulties to find it. To solve this issue, we proposed an approach to convert the optimal path to trajectory. More precisely, a time-based profile trajectory of position and velocity was introduced.

The article is organized as follow: Section 2 describes the problem and case study. Section 3 describes a kinematics model of a mobile robot and the proposed approach, flowcharts and evaluation criteria are introduced in Section 4. The multi objective optimization function is addressed in Section 5 Section 6 is presents the simulation results. In Section 7 conclusions and future work are discussed.

2   Problem Definition

The mobile robot's mission is to travel from a start point to a goal point by minimizing traveling distance and traveling time, maintaining smooth (minimum curvatures) and safety of movement with collision free in 2D complex environment with stationary obstacles.

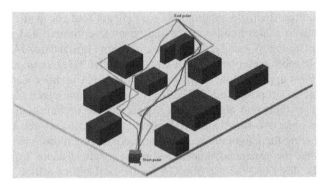

Fig. 1. Problem definition

As shown in Fig.1, even though the shortest path is the red line, it has the lowest security performance. The best smoothness performance path is the orange line, but unfortunately it has the longest path. Although the best security performance is the green path, its length and smoothness are not the best [7, 8]. The trajectory marked in blue has a minimum-time smooth trajectory and minimum energy consumption but the lowest security performance. The multi objective optimal trajectory planning of mobile robot is the dotted black line, a trade off with the advantage that multiple solutions can be obtained in a single run.

3   Path and Trajectory Generating

Automatic trajectory generation is one of the most important functions which an autonomous mobile robot has to be able to move in the environment whilst avoiding obstacles and reaching specific goal configurations as well as saving as much energy as possible [11]. Hence, a trajectory is generated from a geometric path that takes the robot from the initial to its goal position. In order to show a level of intelligence this path must be optimized according to criteria that are important in terms of the robot, working space and given problem [11, 12]. A typical path, however, does not necessarily comply with the robot's particular kinematics and dynamics, such as the non-holonomic constraint of a wheeled mobile robot. Therefore, if actual motion is to be even possible at all, a greater control effort may be required [11]. A trajectory constitutes a time-based profile of position and velocity from start to destination whereas paths are based on non-time parameters. To have smooth movement the parametric Cubic Spline function of trajectory is used, and the trajectory must afford continuous velocity and acceleration. According to that the Kinematics model for mobile robot is described in next section

3.1   Kinematic Model of Mobile Robot

The Kinematic model for mobile robot is shown in Fig.2. The robot has a differential drive system with two independent drive wheels, and a castor wheel for stability. It is assumed that the robot is placed on a plane surface. Hence, the contacts between the wheels of the robot and the rigid horizontal plane have pure rolling and non slipping conditions during the motion. This nonholonomic constraint can be written as [13]:

$$\dot{x} sin\theta - \dot{y} cos\theta = 0 \qquad (1)$$

The center position of mobile robot (x, y, θ) is expressed in the inertial coordinate frame. Here x and y are the position of the robot and is orientation with respect to inertial frame. Supposing that the robot moves on a plane with linear and angular velocities, the state vector can be expressed as $\dot{q} = (\dot{x}, \dot{y}, \dot{\theta})$. The robot's motion on linear trajectories is given by constant velocities of the wheels and the motion on circular trajectories is determined by the difference between the angular velocities of the two drive wheels. The state of the robot is defined by its position and orientation and by the speeds of the mobile robot [13]. A simple structure of a differentially driven three wheeled mobile robot is shown in Fig. 2.

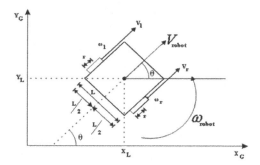

Fig. 2. Kinematics model of Mobile Robot [13]

Path planning under kinematic constraints is transformed into a pure geometric problem. The optimal path is composed of circular arcs and straight lines. Mobile robot motion can be expressed in terms of translational and rotational motion. The translational component is the displacement of the center of the mobile robot and the rotational component is the rotational movement of the mobile robot. The kinematic behaviour of the mobile robot depends on the linear velocity v and the angular velocity ω [13, 14, and 15]. The linear velocities of left and right wheels are v_l and v_r of mobile robot, respectively and can described by (2).

$$v_r = r\omega_r, \quad v_l = r\omega_l \qquad (2)$$

Where ω_r and ω_l are angular velocities of left and right wheels of mobile robot respectively. Both wheels have same radius defined by r. The distance between two wheels is l. For the robot's movement, the linear velocity v and the angular velocity ω are chosen and can be obtain by (3) and (4)

$$\omega = \frac{v_r - v_l}{L} = \omega = \frac{r}{L}(\omega_r - \omega_l) \tag{3}$$

$$, v = \frac{v_r + v_l}{2} = \frac{r}{2}(\omega_r + \omega_l) \tag{4}$$

We can describe the dynamic model for the mobile robot by

$$\begin{bmatrix} \dot{x} \\ \dot{y} \\ \dot{\theta} \end{bmatrix} = \begin{bmatrix} \cos\theta & 0 \\ \sin\theta & 0 \\ 0 & 1 \end{bmatrix} \begin{bmatrix} v \\ \omega \end{bmatrix} \tag{5}$$

Where the tangential velocity in x direction is $\dot{x}$, the tangential velocity in y direction is $\dot{y}$ and the angular velocity of the vehicle is $\dot{\theta}$. This implies

$$x(t) = \int_0^t v(t) \cos(\theta(t)) \, dt \tag{6}$$

$$y(t) = \int_0^t v(t) \sin(\theta(t)) \, dt \tag{7}$$

$$\theta(t) = \int_0^t \omega(t) dt \tag{8}$$

The boundary conditions include the position, velocity and acceleration constraints of the robot [16, 17].

$$x_{initial} = x(t = 0) \ and \ x_{target} = x(t = T) \tag{9}$$

Velocity and acceleration constraints mean that the mobile robot should start from rest at its initial position, accelerate to reach its maximum velocity and near the target location decelerate to stop at the goal position.

4   Proposed Approach

The proposed approach based on performing modification of A* and GA is presented to enhance the searching ability of robot movement towards optimal solution state. In addition, the approach can find a multi objective optimal path and trajectory for mobile robot navigation as well as to use it in complex static environment. The classical method and modified A* search method in initialization stage for single objectives and multi objectives have been proposed to overcome GA drawbacks. Also, in order to avoid fall into a local minimum complex static environment we have proposed several genetic operators such as deletion operator and enhanced mutation operator by adding basic A* to improve the best path partly. The aim of this combination is to enhanced GA efficiency and path planning performance. Hence, several genetic operators are proposed based on domain-specific knowledge and characteristics of path planning to avoid falling into a local minimum in complex environment and to improve the optimal path partly such as deletion operator and enhanced mutation with basic A*. In addition, the proposed approach is received an initial population from a classical method or modified A*. For more details someone could see [10].

The general schema of the proposed approach to the multi objective optimization of path planning and trajectory generating problem can be defined in five main steps as shown in Fig. 4 and flowchart in Fig. 5. First, some of definitions correspond to the initialization stage are presented. We construct 2D static map (indoor area) with and without different numbers of static obstacles. Also, a shortcut or decreased operator has been used to eliminate obstacles nodes from the map at the beginning of the algorithm. The next step is the initial population which is called generating and moving for sub optimal feasible paths. In this stage, the classical method and modified A* are used for generating a set of sub optimal feasible paths in a simple map and a complex map, respectively. Then, the paths obtained are used for establishing the initial population for the GA optimization. Here, the mobile robot moves in an indoor area and it can move in any of eight directions (forward, backward, right, left, right-up, right-down, left-up, and left-down). There are two methods in this step. Someone can use the classical method where the movement of the mobile robot is controlled by a transition rule function, which in turn depends on the Euclidean distance between two points and the roulette wheel method is used to select the next point and to avoid falling in a local minimum on the complex map.

The robot moves through every feasible solution to find the optimal solution in favored tracks that have a relatively short distance between two points, where the location of the mobile robot and the quality of the solution are maintained such that the sub optimal solution can be obtained. When, the number of the obstacles is increasing, the classical method may face difficulties finding a solution or may not even find one. Also, the more via points are used, the more time consuming is the algorithm, as it depends mostly on the number of via points it will use on the path in the complex map. If the modified A* search algorithm is added in the initialization stage of GA proposed approach will find a solution in any case, even if there are many obstacles. The A* algorithm is the most effective free space searching algorithms in term of path length optimization (for a single objective). We propose a modified A* for searching for sub optimal feasible path regardless of length to establish the initial solution of GA in a complex map, by adding the probability function to the A* algorithm. We have modified the A* in order to avoid using the shortest path which could affect the path performance in terms of multi objective (length, security and smoothness) in the initial stage.

$$F(n) = Rand*(g(n) + h(n)). \tag{10}$$

The last step is the modified GA optimization which uses the modified GA for optimizing the search for the sub optimal path that was generated in step above as initial population (chromosomes). Hence, the main stages in the modified GA are natural selection, standard crossover, proposed deletion operator , enhanced mutation with basic A* and sort operator to improve the algorithm's efficiency in terms of path planning, because it is difficult to achieve convergence and generate optimized feasible paths only with operators in the standard GA. A chromosome represents the path and its length varies depending on the case at hand. This means that it consists of a set of genes (via-points) from the start position to the target position of the path. Since $p(x_0, y_0)$ is always the starting point and $p(x_n, y_n)$ is always the target point, the via-points of the path are $p(x_1, y_1)$ and $p(x_{i+1}, y_{i+1})$, and all these points represent the genes of the chromosome as shown in Fig. 3.

$$Path\ (P\ )=\{(x_0, y_0),(x_1, y_1),\dots,(x_{i+1}, y_{i+1}),\dots, (x_n, y_n)\}$$

Fig. 3. Path structure

In each generation, all chromosomes will be evaluated by the fitness function F, which will be discussed later. Thus, a chromosome with minimum fitness has a considerably higher probability than others of selecting and reproducing by means of GA operators in the next generation. The steps from first to fourth are repeated until the maximum number of iterations is reached to find optimal path which it will use to generate the optimal trajectory in step 5 as shown in flowchart Fig. 5. As we mentioned in details in [10] that we proposed enhanced mutation operator and deleting operator. The enhanced mutation operator is enhanced by adding the traditional A* search method to mutation operator and is used to avoid falling into a local minimum, and to improve and decrease the distance of the partial path, between two random points included in the main path. Also, deletion operator is used to delete the repeated points (redundant) from path. For a specific point, the approach reversely checks if this is equal to others, and this is done for each point.

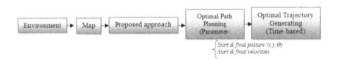

Fig. 4. Proposed approach

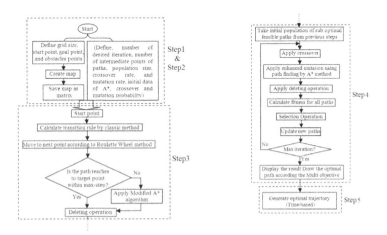

Fig. 5. Flow chart of proposed approach

5   Multi Objective Fitness Function

The idea of Pareto-optimality is introduced to solve multi-objective optimization problem with the advantage that multiple tradeoff solutions can be obtained in a single run [18]. The total cost of fitness (or objective) function of feasible path P with n points is obtained by a linear combination of the weighted sum of multi objectives as follows [7, 8]:

$$\min F(P) = \min \{\omega_1 F_1(P) + \omega_c F_2(P) + \omega_s F_3(P) + \omega_t F_4(P)\} \tag{11}$$

$$F_1(P) = \sum_{i=0}^{n-1} \sqrt{(x_{i+1} - x_i)^2 + (y_{i+1} - y_i)^2} \tag{12}$$

$$F_2(P) = \sum_{i=1}^{n-2} \theta(p_i p_{i+1}, p_{i+1} p_{i+2}) + C_1 \times S. \tag{13}$$

$$F_3(P) = \frac{C_2}{\sum_{i=1}^{n-1} \min\_dist \ (p_i p_{i+1}, OB\ *)} \tag{14}$$

$$F_4(P) = t_T \tag{15}$$

Where w_1, w_s, w_c and w_t represent the weight of each objective to total cost F(P). $F_1(P)$ is the total length of path and criteria of path shortness is defined as the Euclidean distance between two point, $F_2(P)$ is the path smoothness, where $\theta(p_i p_{i+1}, p_{i+1} p_{i+2})$ is the angle between the vectorial path segments $p_i p_{i+1}$ and $p_{i+1} p_{i+2}$, $(0 \le \theta \le \pi)$. C_1 is a positive constant; S is the number of line segments in the path, F_3 (P) is the path clearance or path security, where $\min\_dist(p_i p_{i+1}, OB*)$ is the shortest distance between the path segment $p_i p_{i+1}$ and its proximate obstacle OB *. C_2 is a positive constant and its purpose is to make the numerical scope of $F_3(P)$ in the same order of magnitude with the previous two objective values. $F_4(P)$ represents the total consumed time for robot motion, where t_T are the total time from start to target point. The weights of the shortest, smoothest, security and time fitness functions, w_1, w_s, w_c and w_t respectively, are tuned through simulation and trial and error, with best found values and by minimizing the overall fitness function regarding the assigned weights of each criterion, a suitable optimal path and trajectory can be obtained

6   Simulation Results

The proposal approach has tested in simple and complicated 2D static environments with different numbers of obstacles. MATLAB software (CPU is 2.61 GHz) was used for the simulation. Figs. 6 (a-g), Figs. 7(a-g) and Table 1, 2, 3 show the execution of the program for final Pareto optimized path and trajectory various maps.

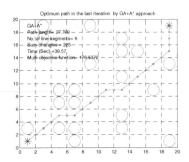

(a):Optimal path planning in Map 1

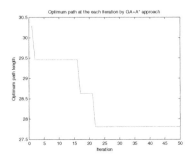

(b): Relationship between optimal path length and iteration in Map 1

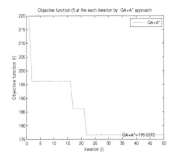

(c): Objective function F versus iteration

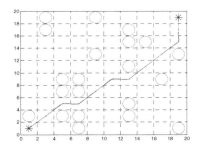

(d): Optimal path and trajectory are generating marked in blue and red , respectively in Map 1

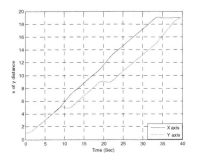

(e): X, Y axes are blue and red, respectively. for optimal trajectory

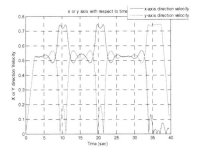

(f): X,Y axes direction velocity for optimal trajectory

Fig. 6. The final Pareto optimized path and trajectory in Map 1

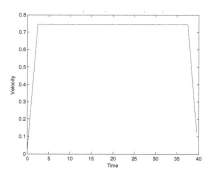

(g): The final velocity profile for the optimal trajectory generation

Fig. 6. (*continued*)

Table 1. statistics results of performance index for proposed approaches in map1

Performance Index	Proposed Approach
Optimal path length	27.799
No. of Segments	6
Sum of angles	225
Travel time (sec.)	39.57
Multi objective value	179.5372
Max. Iteration (i)	50

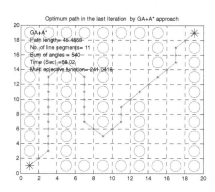

(a):Optimal path planning in Map 2

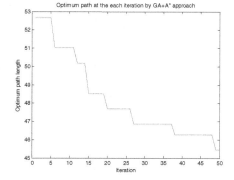

(b): Relationship between optimal path length and iteration and in Map 2

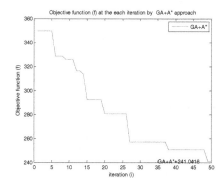

(c): Objective function F versus iteration

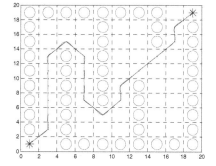

(d): Optimal path and trajectory are generating in blue and red, respectively

Fig. 7. The final Pareto optimized path and trajectory in Map 1

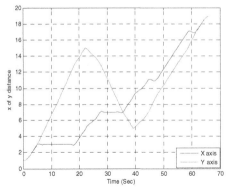

(e): X,Y axis for optimal trajectory

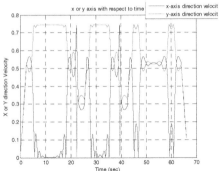

(f): X and Y axes direction velocity are blue and red, respectively. for optimal trajectory

Table 2. Statistics results of performance index for proposed approaches in Map2

Performance Index	Proposed Approach
Optimal path length	45.4558
No. of Segments	11
Sum of angles	540
Travel time (sec.)	66.02
Multi objective value	241.0416

(g): The final velocity profile for the optimal trajectory generation

Fig. 7. (*continued*)

Table 3. Average statistics results of Performance index for proposed approaches of 10 TRAILS in map 2

Performance Index	Proposed Approach
Optimal path length	46.92
No. of Segments	12
Sum of angles	663.75
Travel time (sec.)	70.28
Multi objective value	256.44
Max. Iteration (i)	50

Hence, the proposed approach was tested to generate the optimal collision free path planning and trajectory generation in terms of length, smoothness, security and time in complex static environment. The results for the multi objective optimal path and trajectory for the robot are shown in Fig 6 (a-d) and 7 (a-d), respectively. Hence, the statistics results of performance index for proposed approach in Tables (1-3) Figs. 6(e) and Figs. 7(e) show the X and Z coordinate with time for optimal trajectory in map 1 and 2, respectively. As shown in Figs. 6 (f) and Figs. 7(f) the velocity profile in

X and Y direction for the mobile robot in Map1 and 2, respectively. The final velocity profile for the optimal trajectory generation in Fig. 6 (g) and Fig. 7(g). The simulation results show that the mobile robot travels successfully from one location to another and reaches its goal after avoiding all obstacles that are located in its way in all tested environments, and indicates that the proposed approach is accurate and can find a set Pareto optimal solution efficiently in a single run as shown in Fig. 8.

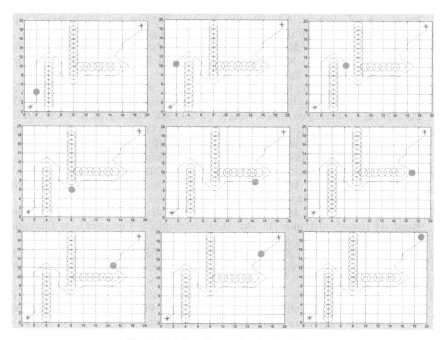

Fig. 8. Mobile robot navigation in Map 3

7 Conclusion

In this work, we presented a proposed approach to generate multi objective optimization of a path and trajectory of mobile robot with collision free in static environment as well as to enhance the searching ability of robot movement towards optimal solution state in static. The simulation results show that the proposed approach is able to achieve multi objective optimization in static environment efficiently. Also, it has the ability to find a solution when the number of obstacles is increasing. Extensions of this work could take into account additional costs, such as avoid obstacles in dynamic environments.

Acknowledgments. The authors gratefully acknowledge the German Academic Exchange Service (DAAD), the University of Siegen and the Iraqi Ministry of Higher Education and Scientific Research for financial support.

References

[1] Krishnan, P.S., Paw, J.K.S., Tiong, S.K.: Cognitive Map Approach for Mobility Path Optimization using Multiple Objectives Genetic Algorithm. In: 4th IEEE International Conference on Autonomous Robots and Agents, pp. 267–272 (2009)

[2] Yongnian, Z., Yongping, L., Lifang, Z.: An Improved Genetic Algorithm for Mobile Robotic Path Planning. In: 24th IEEE Conference on Control and Decision Conference, pp. 3255–3260 (2012)

[3] Panteleimon, Z., Ijspeert, A.J., Degallier, S.: Path Planning with the humanoid robot iCub. Semester project, Biologically inspired Robotics Group, Birg (2009)

[4] Castillo, O., Trujillo, L.: Multiple Objective Optimization Genetic Algorithms for Path Planning in Autonomous Mobile Robots. International Journal of Computers, Systems and Signals 6(1), 48–63 (2005)

[5] Purcaru, C., Precup, R.E., Iercan, D., Fedorovici, L.O., David, R.C.: Hybrid PSO-GSA Robot Path Planning Algorithm in Static Environments with Danger Zones. In: 17th IEEE International Conference on System Theory, Control and Computing, pp. 434–439 (2013)

[6] Solteiro Pires, E.J., Tenreiro Machado, J.A., De Moura Oliveira, P.B.: Robot Trajectory Planning Using Multi-objective Genetic Algorithm Optimization. In: Deb, K., Tari, Z. (eds.) GECCO 2004. LNCS, vol. 3102, pp. 615–626. Springer, Heidelberg (2004)

[7] Jun, H., Qingbao, Q.: Multi-Objective Mobile Robot Path Planning based on Improved Genetic Algorithm. In: IEEE International Conference on Intelligent Computation Technology and Automation, vol. 2, pp. 752–756 (2010)

[8] Geetha, S., Chitra, G.M., Jayalakshmi, V.: Multi Objective Mobile Robot Path Planning based on Hybrid Algorithm. In: 3rd IEEE International Conference on Electronics Computer Technology, vol. 6, pp. 251–255 (2011)

[9] Gong, D.W., Zhang, J.H., Zhang, Y.: Multi-Objective Particle Swarm Optimization for Robot Path Planning in Environment with Danger Sources. Journal of Computers 6(8), 1554–1561 (2011)

[10] Oleiwi, B.K., Hubert, R., Kazem, B.: Modified Genetic Algorithm based on A* algorithm of Multi objective optimization for Path Planning. In: 6th International Conference on Computer and Automation Engineering, vol. 2(4), pp. 357–362 (2014)

[11] Trajano, T.A.A., Armando, A.M.F., Max, M.S.D.: Parametric Trajectory Generation for Mobile Robots. ABCM Symposium Series in Mechatronics, vol. 3, pp. 300–307 (2008)

[12] Sedaghat, N.: Mobile Robot Path Planning by New Structured Multi-Objective Genetic Algorithm. In: IEEE International Conference on Soft Computing and Pattern Recognition, pp. 79–83 (2011)

[13] Vivekananthan, R., Karunamoorthy, L.: A Time Optimal Path Planning for Trajectory Tracking of Wheeled Mobile Robots. Journal of Automation, Mobile Robotics & Intelligent Systems 5(2), 35–41 (2011)

[14] Alves, S.F.R., Rosario, J.M., Filho, H.F., Rincon, L.K.A., Yamasaki, R.A.T.: Conceptual Bases of Robot Navigation Modeling Control and Applications. Alejandra Barrera (2011)

[15] Xianhua, J., Motai, Y., Zhu, X.: Predictive fuzzy control for a mobile robot with non-holonomic constraints. In: IEEE Mid-Summer Workshop on Soft Computing in Industrial Applications, Helsinki University of Technology, Espoo, Finland (2005)

[16] Breyak, M., Petrovic, I.: Time Optimal Trajectory Planning Along Predefined Path for Mobile Robots with Velocity and Acceleration Constraints. In: IEEE/ASME International Conference on Advanced Intelligent Mechatronics, pp. 942–947 (2011)

[17] Arshad, M., Choudhry, M.A.: Trajectory Planning of Mobile robot in Unstructured Environment for Multiple Objects. Mehran University Research Journal of Engineering & Technology 31(1), 39–50 (2012)

[18] Fonseca, C.M., Fleming, P.J.: An Overview of Evolutionary Algorithms in Multi-objective Optimization. Evolutionary Computing 3(1), 1–16 (1995)

New Procedures of Pattern Classification
for Vibration-Based Diagnostics via Neural Network

Nicholas Nechval[1], Konstantin Nechval[2], and Irina Bausova[1]

[1] University of Latvia, EVF Research Institute, Statistics Department,
Raina Blvd 19, LV-1050 Riga, Latvia
Nicholas Nechval, Irina Bausova
nechval@junik.lv
[2] Transport and Telecommunication Institute, Applied Mathematics Department,
Lomonosov Street 1, LV-1019 Riga, Latvia
konstan@tsi.lv

Abstract. In this paper, the new distance-based embedding procedures of pattern classification for vibration-based diagnostics of gas turbine engines via neural network are proposed. Diagnostics of gas turbine engines is important because of the high cost of engine failure and the possible loss of human life. Engine monitoring is performed using either 'on-line' systems, mounted within the aircraft, that perform analysis of engine data during flight, or 'off-line' ground-based systems, to which engine data is downloaded from the aircraft at the end of a flight. Typically, the health of a rotating system such as a gas turbine is manifested by its vibration level. Efficiency of gas turbine monitoring systems primarily depends on the accuracy of employed algorithms, in particular, pattern recognition techniques to diagnose gas path faults. For pattern recognition of vibration signals, the characteristics usually used are: (1) amplitude, (2) frequency, and (3) space. In investigations, many techniques were applied to recognize gas path faults, but recommendations on selecting the best technique for real monitoring systems are still insufficient and often contradictory. In this paper, the new distance-based embedding procedures for pattern classification (recognition) are presented. These procedures do not require the arbitrary selection of priors as in the Bayesian classifier and allow one to take into account the cases which are not adequate for Fisher's Linear Discriminant Analysis (FLDA). The results obtained in this paper agree with the simulation results, which confirm the validity of the theoretical predictions of performance of the presented procedures. The computer simulation results are promising.

Keywords: Engine, diagnostics, features, pattern classification, fault detection.

1   Introduction

The machines and structural components require continuous monitoring for the detection of fatigue cracks and crack growth for ensuring an uninterrupted service. Non-destructive testing methods like ultrasonic testing, X-ray, etc., are generally

V. Golovko and A. Imada (Eds.): ICNNAI 2014, CCIS 440, pp. 63–75, 2014.
© Springer International Publishing Switzerland 2014

useful for the purpose. These methods are costly and time consuming for long components, e.g., railway tracks, long pipelines, etc. Vibration-based methods can offer advantages in such cases [1]. This is because measurement of vibration parameters like natural frequencies is easy. Further, this type of data can be easily collected from a single point of the component. This factor lends some advantages for components, which are not fully accessible. This also helps to do away with the collection of experimental data from a number of data points on a component, which is involved in a prediction based on, for example, mode shapes. Nondestructive evaluation (NDE) of structures using vibration for early detection of cracks has gained popularity over the years and, in the last decade in particular, substantial progress has been made in that direction. Almost all crack diagnosis algorithms based on dynamic behaviour call for a reference signature. The latter is measured on an identical uncracked structure or on the same structure at an earlier stage. Dynamics of cracked rotors has been a subject of great interest for the last three decades and detection and monitoring have gained increasing importance, recently. Failures of any high speed rotating components (jet engine rotors, centrifuges, high speed fans, etc.) can be very dangerous to surrounding equipment and personnel (see Fig. 1), and must always be avoided.

Fig. 1. Jet engine fan section failure

Jet engine disks operate under high centrifugal and thermal stresses. These stresses cause microscopic damage as a result of each flight cycle as the engine starts from the cold state, accelerates to maximum speed for take-off, remains at speed for cruise, then spools down after landing and taxi. The cumulative effect of this damage over time creates a crack at a location where high stress and a minor defect combine to create a failure initiation point. As each flight operation occurs, the crack is enlarged by an incremental distance. If allowed to continue to a critical dimension, the crack would eventually cause the burst of the disk and lead to catastrophic failure (burst) of the engine. Engine burst in flight is rarely survivable.

In this paper, we will focus on aircraft or jet engines, which are a special class of gas turbine engines. Typically, physical faults in a gas turbine engine include problems such as erosion, corrosion, fouling, built-up dirt, foreign object damage (FOD), worn seals, burned or bowed blades, etc. These physical faults can occur individually or in combination and cause changes in performance characteristics of the compressors, and in their expansion and compression efficiencies. In addition, the faults cause changes in the turbine and exhaust system nozzle areas. These changes in the performance of the gas turbine components result in changes in the measurement parameters, which are therefore dependent variables.

2   Vibration-Based Diagnostics: Problem Statement

In this section, we look at a problem where vibration characteristics are used for gas turbine diagnostics. The present chapter focuses on turbine blade damage. Turbine blades undergo cyclic loading causing structural deterioration, which can lead to failure. It is important to know how much damage has taken place at any particular time to monitor the condition or health of the blade and to avoid any catastrophic failure of the blades. Several studies look at damage at a given time during the operational history of the structure. This is typically called diagnostics and involves detection, location, and isolation of damage from a set of measured variables. The detection function is most fundamental, as it points out if the damage is present or not. However, some level of damage due to microcracks and other defects is always present in a structure. The important issue of indicating when to detect damage depends on how much life is left in the structure. It is not advantageous to detect small levels of damage in a structure. It would be useful if damage detection were triggered some time before final failure. The subject of prognostics involves predicting the evolution of structural or vibrational characteristics of the system with time and is important for prediction of failure due to operational deterioration. Some recent studies have considered dynamical systems approaches to model damage growth based on differential equations [2], while others have used physics-based models [3]. The stiffness of the structure is gradually reduced with crack growth, and stiffness is related to the vibrational characteristics of the structure. The decreased frequency shows that stiffness of the structure is decreasing, and thus serves as a damage indicator for monitoring crack growth in the structure. Selected studies have looked at modeling turbine blades as rotating Timoshenko beams with twist and taper [4–6]. Some studies have addressed damage in such beams using vibrational characteristics [7, 8]. However, these studies typically address damage at a given time point in the operational history and do not look at the effect of damage growth on the vibrational characteristics. In additions, turbine blades are designed to sustain a considerable amount of accumulated damage prior to failure. Therefore, it is desirable to indicate that a blade is damaged at the point when its operational life is almost over.

The problem of vibration-based diagnostics consists in the following. There are m classes (populations) of vibration signal, the elements (vibrational characteristics) of which are characterized by p measurements (features). Next, suppose that we are investigating a target vibration signal on the basis of the corresponding p measurements. We postulate that this signal can be regarded as a 'random drawing' from one of the m populations (classes) but we do not know from which one. We suppose that m samples are available, each sample being drawn from a different class.

The elements of these samples are realizations of p-dimensional random variables. After a sample of p-dimensional vectors of measurements of vibrational characteristics of the signal is drawn from a class known a priori to be one of the above set of m classes, the problem is to infer from which class the sample has been drawn. The decision rule should be in the form of associating the sample of observations on the target vibration signal with one of the m samples and declaring that this signal has come from the same class as the sample with which it is associated. Fig. 2 shows the structure of the proposed system for vibration-based diagnostics.

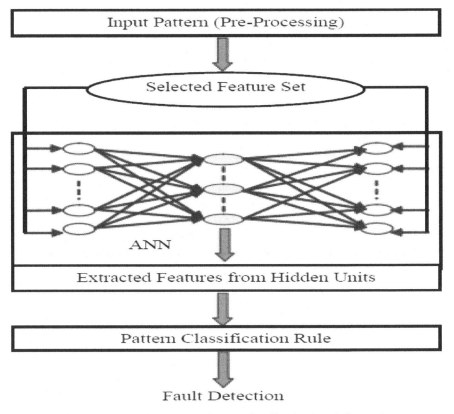

Fig. 2. Structure of the proposed system for vibration-based diagnostics

3   Classification Procedures for Vibration-Based Diagnostics

Classification is often referred to simply as discriminant analysis. In engineering and computer science, classification is usually called pattern recognition. Some writers use the term classification analysis to describe cluster analysis, in which the observations are clustered according to variable values rather than into predefined classes. In classification, a sampling unit (subject or object) whose class membership is unknown is assigned to a class on the basis of the vector of p measured values, $\mathbf{y}$, associated with the unit. To classify the unit, we must have available a previously obtained sample of observation vectors from each class. Then one approach is to compare $\mathbf{y}$ with the mean vectors $\overline{\mathbf{y}}_1, \overline{\mathbf{y}}_2, ..., \overline{\mathbf{y}}_k$ of the k samples and assign the unit to the class whose $\overline{\mathbf{y}}_i$ is closest to $\mathbf{y}$.

3.1   Fisher's Procedure for Pattern Classification into Two Classes

When there are two populations (classes), we can use a classification procedure due to Fisher [9]. The principal assumption for Fisher's procedure is that the two populations have the same covariance matrix ($\Sigma_1 = \Sigma_2$). Normality is not required. We obtain a sample from each of the two populations and compute $\overline{\mathbf{y}}_1, \overline{\mathbf{y}}_2$, and $\mathbf{S}_{12}$. A simple procedure for classification into one of the two classes denoted by C_1 and C_2 can be based on the discriminant function,

$$z = \mathbf{w}'\mathbf{y} = (\overline{\mathbf{y}}_1 - \overline{\mathbf{y}}_2)'\mathbf{S}_{12}^{-1}\mathbf{y}, \tag{1}$$

where $\mathbf{y}$ is the vector of measurements on a new sampling unit that we wish to classify into one of the two classes (populations), $\mathbf{w}$ is a direction which is determined from maximization of the ratio of between-class to within-class variances proposed by Fisher,

$$J_F = \frac{[\mathbf{w}'(\overline{\mathbf{y}}_1 - \overline{\mathbf{y}}_2)]^2}{\mathbf{w}'\mathbf{S}_{12}\mathbf{w}}, \tag{2}$$

$\mathbf{S}_{12}$ is the pooled within-class covariance matrix, in its bias-corrected form given by

$$\mathbf{S}_{12} = \frac{(n_1 - 1)\mathbf{S}_1 + (n_2 - 1)\mathbf{S}_2}{n_1 + n_2 - 2}, \tag{3}$$

$\mathbf{S}_1$ and $\mathbf{S}_2$ are the unbiased estimates of the covariance matrices of classes C_1 and C_2, respectively, and there are n_i observations in class C_i ($n_1+n_2=n$). The solution for $\mathbf{w}$ that maximizes J_F can be obtained by differentiating J_F with respect to $\mathbf{w}$ and equating to zero. This yields

$$\frac{2\mathbf{w}'(\overline{\mathbf{y}}_1 - \overline{\mathbf{y}}_2)}{\mathbf{w}'\mathbf{S}_{12}\mathbf{w}} \left[(\overline{\mathbf{y}}_1 - \overline{\mathbf{y}}_2) - \left(\frac{\mathbf{w}'(\overline{\mathbf{y}}_1 - \overline{\mathbf{y}}_2)}{\mathbf{w}'\mathbf{S}_{12}\mathbf{w}} \right) \mathbf{S}_{12}\mathbf{w} \right] = 0. \tag{4}$$

Since we are interested in the direction of $\mathbf{w}$ (and noting that $\mathbf{w}'(\overline{\mathbf{y}}_1 - \overline{\mathbf{y}}_2)/\mathbf{w}'\mathbf{S}_{12}\mathbf{w}$ is a scalar), we must have

$$\mathbf{w} \propto \mathbf{S}_{12}^{-1}(\overline{\mathbf{y}}_1 - \overline{\mathbf{y}}_2). \tag{5}$$

We may take equality without loss of generality. For convenience we speak of classifying $\mathbf{y}$ rather than classifying the subject or object associated with $\mathbf{y}$.

To determine whether $\mathbf{y}$ is closer to $\overline{\mathbf{y}}_1$ or $\overline{\mathbf{y}}_2$, we check to see if z in (1) is closer to the transformed mean $\overline{z}_1$ or to $\overline{z}_2$, where

$$\overline{z}_1 = \mathbf{w}'\overline{\mathbf{y}}_1 = (\overline{\mathbf{y}}_1 - \overline{\mathbf{y}}_2)'\mathbf{S}_{12}^{-1}\overline{\mathbf{y}}_1, \tag{6}$$

$$\overline{z}_2 = \mathbf{w}'\overline{\mathbf{y}}_2 = (\overline{\mathbf{y}}_1 - \overline{\mathbf{y}}_2)'\mathbf{S}_{12}^{-1}\overline{\mathbf{y}}_2. \tag{7}$$

Fisher's linear classification procedure [9] assigns $\mathbf{y}$ to C_1 if $z = \mathbf{w}'\mathbf{y}$ is closer to $\overline{z}_1$ than to $\overline{z}_2$ and assigns $\mathbf{y}$ to C_2 if z is closer to $\overline{z}_2$. It will be noted that z is closer to $\overline{z}_1$ if

$$z > \frac{\overline{z}_1 + \overline{z}_2}{2}. \tag{8}$$

This is true in general because $\overline{z}_1$ is always greater than $\overline{z}_2$, which can easily be shown as follows:

$$\overline{z}_1 - \overline{z}_2 = \mathbf{w}'(\overline{\mathbf{y}}_1 - \overline{\mathbf{y}}_2) = (\overline{\mathbf{y}}_1 - \overline{\mathbf{y}}_2)'\mathbf{S}_{12}^{-1}(\overline{\mathbf{y}}_1 - \overline{\mathbf{y}}_2) > 0, \tag{9}$$

because $\mathbf{S}_{12}^{-1}$ is positive definite. Thus $\overline{z}_1 > \overline{z}_2$. [If $\mathbf{w}$ were of the form $\mathbf{w}' = (\overline{\mathbf{y}}_2 - \overline{\mathbf{y}}_1)'\mathbf{S}_{12}^{-1}$, then $\overline{z}_2 - \overline{z}_1$ would be positive.] Since $(\overline{z}_1 + \overline{z}_2)/2$ is the midpoint, $z > (\overline{z}_1 + \overline{z}_2)/2$ implies that z is closer to $\overline{z}_1$. By (9) the distance from $\overline{z}_1$ to $\overline{z}_2$ is the same as that from $\overline{\mathbf{y}}_1$ to $\overline{\mathbf{y}}_2$.

To express the classification rule in terms of $\mathbf{y}$, we first write $(\overline{z}_1 + \overline{z}_2)/2$ in the form

$$\frac{\overline{z}_1 + \overline{z}_2}{2} = \frac{\mathbf{w}'(\overline{\mathbf{y}}_1 + \overline{\mathbf{y}}_2)}{2} = \frac{(\overline{\mathbf{y}}_1 - \overline{\mathbf{y}}_2)'\mathbf{S}_{12}^{-1}(\overline{\mathbf{y}}_1 + \overline{\mathbf{y}}_2)}{2}. \tag{10}$$

Then the classification rule becomes: Assign $\mathbf{y}$ to C_1 if

$$\mathbf{w'y} = (\overline{\mathbf{y}}_1 - \overline{\mathbf{y}}_2)\mathbf{S}_{12}^{-1}\mathbf{y} > \frac{(\overline{\mathbf{y}}_1 - \overline{\mathbf{y}}_2)'\mathbf{S}_{12}^{-1}(\overline{\mathbf{y}}_1 + \overline{\mathbf{y}}_2)}{2}, \tag{11}$$

and assign $\mathbf{y}$ to C_2 if

$$\mathbf{w'y} = (\overline{\mathbf{y}}_1 - \overline{\mathbf{y}}_2)\mathbf{S}_{12}^{-1}\mathbf{y} < \frac{(\overline{\mathbf{y}}_1 - \overline{\mathbf{y}}_2)'\mathbf{S}_{12}^{-1}(\overline{\mathbf{y}}_1 + \overline{\mathbf{y}}_2)}{2}. \tag{12}$$

Fisher's approach [9] using (11) and (12) is essentially nonparametric because no distributional assumptions were made. However, if the two populations are normal with equal covariance matrices, then this method is (asymptotically) optimal; that is, the probability of misclassification is minimized.

3.2   Embedding Procedures for Pattern Classification into Several Classes

Classification via Total Mahalanobis Distance. Let us assume that each of the k populations has the same covariance matrix ($\Sigma_1 = \Sigma_2 = \cdots = \Sigma_k$). The Mahalanobis distance between two vectors $\overline{\mathbf{y}}_i$ and $\overline{\mathbf{y}}_j$, where $i, j \in \{1, 2, ..., k\}$, $i \neq j$, is given by

$$d_{ij} = (\overline{\mathbf{y}}_i - \overline{\mathbf{y}}_j)'\mathbf{S}_{ij}^{-1}(\overline{\mathbf{y}}_i - \overline{\mathbf{y}}_j). \tag{13}$$

If $\mathbf{y}$ has been embedded in the sample from C_i, the Mahalanobis distance between two vectors $\overline{\mathbf{y}}_{\bullet i}$ and $\overline{\mathbf{y}}_j$ is given by

$$d_{\bullet ij} = (\overline{\mathbf{y}}_{\bullet i} - \overline{\mathbf{y}}_j)'\mathbf{S}_{\bullet ij}^{-1}(\overline{\mathbf{y}}_{\bullet i} - \overline{\mathbf{y}}_j). \tag{14}$$

If $\mathbf{y}$ has been embedded in the sample from C_j, the Mahalanobis distance between two vectors $\overline{\mathbf{y}}_i$ and $\overline{\mathbf{y}}_{j\bullet}$ is given by

$$d_{ij\bullet} = (\overline{\mathbf{y}}_i - \overline{\mathbf{y}}_{j\bullet})'\mathbf{S}_{ij\bullet}^{-1}(\overline{\mathbf{y}}_i - \overline{\mathbf{y}}_{j\bullet}). \tag{15}$$

Let

$$d_r(\mathbf{y}) = \sum_{i=1}^{k-1} \sum_{j=i+1}^{k} d_{ij}, \quad r \in \{1, 2, ..., k\} \tag{16}$$

be the total Mahalanobis distance in the case of pattern classification into k classes, where

$$d_{ij} = d_{\bullet ij}, \quad \text{if } i = r, \tag{17}$$

and

$$d_{ij} = d_{ij_\bullet}, \quad \text{if } j = r. \tag{18}$$

Then the classification rule becomes: Assign $\mathbf{y}$ to the class C_r, $r \in \{1, 2, \ldots, k\}$, for which $d_r(\mathbf{y})$ is largest.

If $(\Sigma_1 = \Sigma_2 = \cdots = \Sigma_k)$ does not hold, then instead of the pooled sample covariance matrix

$$\mathbf{S}_{ij} = \frac{(n_i - 1)\mathbf{S}_i + (n_j - 1)\mathbf{S}_j}{n_i + n_j - 2} \tag{19}$$

we use

$$\mathbf{S}_{ij}^{\circ} = \frac{\mathbf{S}_i}{n_i} + \frac{\mathbf{S}_j}{n_j}. \tag{20}$$

Classification via Total Generalized Euclidean Distance. Let us assume that each of the k populations has the same covariance matrix $(\Sigma_1 = \Sigma_2 = \cdots = \Sigma_k)$. We can estimate the common population covariance matrix by a pooled sample covariance matrix

$$\mathbf{S}_{\mathrm{pl}} = \sum_{i=1}^{k} (n_i - 1)\mathbf{S}_i \left[\sum_{i=1}^{k} n_i - k \right]^{-1}, \tag{21}$$

where n_i and $\mathbf{S}_i$ are the sample size and covariance matrix of the ith class. The generalized Euclidean distance between two vectors $\overline{\mathbf{y}}_i$ and $\overline{\mathbf{y}}_j$, where $i, j \in \{1, 2, \ldots, k\}$, $i \neq j$, is given by

$$\tilde{d}_{ij} = \frac{(\overline{\mathbf{y}}_i - \overline{\mathbf{y}}_j)'(\overline{\mathbf{y}}_i - \overline{\mathbf{y}}_j)}{|\mathbf{S}_{\mathrm{pl}}|}, \tag{22}$$

If $\mathbf{y}$ has been embedded in the sample from C_i, then the generalized Euclidean distance between two vectors $\overline{\mathbf{y}}_{\bullet i}$ and $\overline{\mathbf{y}}_j$ is given by

$$\tilde{d}_{\bullet ij} = \frac{(\overline{\mathbf{y}}_{\bullet i} - \overline{\mathbf{y}}_j)'(\overline{\mathbf{y}}_{\bullet i} - \overline{\mathbf{y}}_j)}{|\mathbf{S}_{\mathrm{pl}(\bullet i)}|}. \tag{23}$$

If $\mathbf{y}$ has been embedded in the sample from C_j, then the generalized Euclidean distance between two vectors $\overline{\mathbf{y}}_i$ and $\overline{\mathbf{y}}_{j\bullet}$ is given by

$$\tilde{d}_{ij\bullet} = \frac{(\overline{\mathbf{y}}_i - \overline{\mathbf{y}}_{j\bullet})'(\overline{\mathbf{y}}_i - \overline{\mathbf{y}}_{j\bullet})}{|\mathbf{S}_{\mathrm{pl}(j\bullet)}|}. \tag{24}$$

Let

$$\tilde{d}_r(\mathbf{y}) = \sum_{i=1}^{k-1}\sum_{j=i+1}^{k} \tilde{d}_{ij}, \quad r \in \{1, 2, ..., k\}, \tag{25}$$

be the total generalized Euclidean distance in the case of pattern classification into k classes, where

$$\tilde{d}_{ij} = \tilde{d}_{\bullet ij}, \quad \text{if } i = r, \tag{26}$$

and

$$\tilde{d}_{ij} = \tilde{d}_{ij\bullet}, \quad \text{if } j = r. \tag{27}$$

Then the classification rule becomes: Assign $\mathbf{y}$ to the class C_r, $r \in \{1, 2, ..., k\}$, for which $\tilde{d}_r(\mathbf{y})$ is largest.

If $(\Sigma_1 = \Sigma_2 = \cdots = \Sigma_k)$ does not hold, then instead of $\mathbf{S}_{\mathrm{pl}}$ we use

$$\mathbf{S}^\circ = \sum_{i=1}^{k} \frac{\mathbf{S}_i}{n_i}. \tag{28}$$

Classification via Total Modified Euclidean Distance. Let us assume that each of the k populations has the same covariance matrix $(\Sigma_1 = \Sigma_2 = \cdots = \Sigma_k)$. The modified Euclidean distance between two vectors $\overline{\mathbf{y}}_i$ and $\overline{\mathbf{y}}$, $i \in \{1, 2, ..., k\}$, is given by

$$\breve{d}_i = \frac{(\overline{\mathbf{y}}_i - \overline{\mathbf{y}})'(\overline{\mathbf{y}}_i - \overline{\mathbf{y}})}{|\mathbf{S}_{\mathrm{pl}}|}, \tag{29}$$

where

$$\overline{\mathbf{y}} = \sum_{i=1}^{k} n_i \overline{\mathbf{y}}_i \bigg/ \sum_{i=1}^{k} n_i \tag{30}$$

represents the 'overall average'. If $\mathbf{y}$ has been embedded in the sample from C_i, then the modified Euclidean distance between two vectors $\overline{\mathbf{y}}_{\bullet i}$ and $\overline{\mathbf{y}}$ is given by

$$\breve{d}_{\bullet i} = \frac{(\overline{\mathbf{y}}_{\bullet i} - \overline{\mathbf{y}})'(\overline{\mathbf{y}}_{\bullet i} - \overline{\mathbf{y}})}{|\mathbf{S}_{\mathrm{pl}(\bullet i)}|}, \quad i \in \{1, 2, ..., k\}. \tag{31}$$

Let

$$\breve{d}_r(\mathbf{y}) = \sum_{i=1}^{k} \breve{d}_i, \quad r \in \{1, 2, ..., k\}, \tag{32}$$

be the total modified Euclidean distance in the case of pattern classification into k classes, where

$$\breve{d}_i = \breve{d}_{\bullet i}, \text{ if } i = r. \tag{33}$$

Then the classification rule becomes: Assign $\mathbf{y}$ to to the class C_r, $r \in \{1, 2, ..., k\}$, for which $\breve{d}_r(\mathbf{y})$ is largest.

If $(\Sigma_1 = \Sigma_2 = \cdots = \Sigma_k)$ does not hold, then instead of $\mathbf{S}_{pl}$ we use $\mathbf{S}^{\circ}$ (28).

4   Illustrative Example of Pattern Classification

Consider the observations on $p=2$ variables from $k=3$ populations (classes) [10]. The input data samples are given below.

$$(n_1=3) \qquad\qquad (n_2=3) \qquad\qquad (n_3=3)$$

$$C_1 = \begin{bmatrix} -2 & 5 \\ 0 & 3 \\ -1 & 1 \end{bmatrix}; \quad C_2 = \begin{bmatrix} 0 & 6 \\ 2 & 4 \\ 1 & 2 \end{bmatrix}; \quad C_3 = \begin{bmatrix} 1 & -2 \\ 0 & 0 \\ -1 & -4 \end{bmatrix}. \tag{34}$$

We found that

$$\bar{\mathbf{y}}_1 = \begin{bmatrix} -1 \\ 3 \end{bmatrix}, \quad \bar{\mathbf{y}}_2 = \begin{bmatrix} 1 \\ 4 \end{bmatrix}, \quad \bar{\mathbf{y}}_3 = \begin{bmatrix} 0 \\ -2 \end{bmatrix}, \quad \bar{\mathbf{y}} = \begin{bmatrix} 0 \\ 5/3 \end{bmatrix}. \tag{35}$$

$$\mathbf{S}_{pl} = \begin{bmatrix} 1 & -0.33333 \\ -0.33333 & 4 \end{bmatrix}. \tag{36}$$

Suppose that we have to classify the new observation $\mathbf{y}' = [1, 3]$ into the above classes. Let us assume that each of the $k=3$ populations has the same covariance matrix ($\Sigma_1 = \Sigma_2 = \Sigma_3$).

Classification via Total Mahalanobis Distance. It follows from (16) that

$$d_1(\mathbf{y}) = 20.86, \quad d_2(\mathbf{y}) = 26.33, \quad d_3(\mathbf{y}) = 15.49. \tag{37}$$

Thus, since

$$\tilde{d}_2(\mathbf{y}) = \max_r \tilde{d}_r(\mathbf{y}), \tag{38}$$

we assign $\mathbf{y}$ to class C_2.

Classification via Total Generalized Euclidean Distance. It follows from (25) that

$$\tilde{d}_1(\mathbf{y}) = 15.14, \quad \tilde{d}_2(\mathbf{y}) = 21.91, \quad \tilde{d}_3(\mathbf{y}) = 7.51. \tag{39}$$

Thus, since

$$\tilde{d}_2(\mathbf{y}) = \max_r \tilde{d}_r(\mathbf{y}), \tag{40}$$

we assign $\mathbf{y}$ to class C_2.

Classification via Total Modified Euclidean Distance. It follows from (32) that

$$\breve{d}_1(\mathbf{y}) = 5.066, \quad \breve{d}_2(\mathbf{y}) = 7.312, \quad \breve{d}_3(\mathbf{y}) = 2.596. \tag{41}$$

Thus, since

$$\breve{d}_2(\mathbf{y}) = \max_r \breve{d}_r(\mathbf{y}), \tag{42}$$

we assign $\mathbf{y}$ to class C_2.

It will be noted that the procedures proposed in this paper give the same result in the above case that of Fisher's procedure which was used in [10].

5   Conclusion and Future Work

Since the fault diagnosis problem can be considered as a multi-class classification problem, pattern recognition methods with good generalization and accurate performances have been proposed in recent years. Choi et al. [11] proposed a fault detection and isolation methodology based on principal component analysis–Gaussian mixture model and discriminant analysis–Gaussian mixture model. Fisher linear discriminant analysis (FLDA) has been proved to outperform PCA in discriminating different classes, in the aspect that PCA aims at reconstruction instead of classification, while FLDA seeks directions that are optimal for discrimination [12]. Fisher's linear discriminant analysis is a widely used multivariate statistical technique with two closely related goals: discrimination and classification. The technique is very popular among users of discriminant analysis. Some of the reasons for this are its simplicity and unnecessity of strict assumptions. In its original form, proposed by Fisher, the method assumes equality of population covariance matrices, but does not explicitly require multivariate normality. However, optimal classification performance of Fisher's discriminant function can only be expected when multivariate normality is present as well, since only good discrimination can ensure good allocation. In practice, we often are in need of analyzing input data samples, which are not adequate for Fisher's classification rule, such that the distributions of the groups are not multivariate normal or covariance matrices of those are different or there are strong multi-nonlinearities. An example of one of situations of pattern classification, which produces non-linear separation of classes and is not adequate for Fisher's classification rule, would become clear from the illustration shown in Fig. 3.

Separation may be easier in higher dimensions

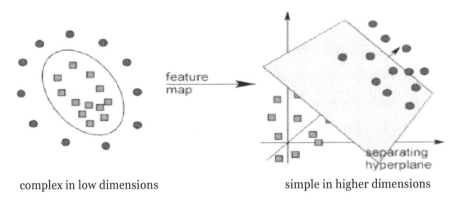

complex in low dimensions   simple in higher dimensions

Fig. 3. Pattern classification situation which is not adequate for FLDA

One solution to address this problem would be the application of kernels to input data, which would essentially transform the input to a higher dimensional space, wherein the probability of linearly separating the classes is higher.

This paper proposes the improved approaches to pattern classification for vibration-based diagnostics via neural network which represent the new distance-based embedding procedures that allow one to take into account the cases which are not adequate for Fisher's classification rule. Moreover, these approaches allow one to classify sets of multivariate observations, where each of the sets contains more than one observation. For the cases, which are adequate for Fisher's classification rule, the proposed approaches give the results similar to that of FLDA.

The methodology described here can be extended in several different directions to handle various problems of pattern classification (recognition) that arise in practice (in particular, the problem of changepoint detection in a sequence of multivariate observations).

Acknowledgments. This research was supported in part by Grant No. 06.1936, Grant No. 07.2036, and Grant No. 09.1014 from the Latvian Council of Science and the National Institute of Mathematics and Informatics of Latvia.

References

1. Dimarogonas, A.D.: Vibration of Cracked Structures: a State of the Art Review. Engineering and Fracture Mechanics 55, 831–857 (1996)
2. Adams, D.E., Nataraju, M.: A Nonlinear Dynamics System for Structural Diagnosis and Prognosis. International Journal of Engineering Science 40, 1919–1941 (2002)
3. Roy, N., Ganguli, R.: Helicopter Rotor Blade Frequency Evolution with Damage Growth and Signal Processing. Journal of Sound and Vibration 283, 821–851 (2005)
4. Krupka, R.M., Baumanis, A.M.: Bending-Bending Mode of Rotating Tapered Twisted Turbo Machine Blades Including Rotary Inertia and Shear Deflection. Journal of Engineering for Industry 91, 10–17 (1965)
5. Thomas, J., Abbas, B.A.H.: Finite Element Model for Dynamic Analysis of Timoshenko Beam. Journal of Sound and Vibration 41, 291–299 (1975)
6. Rao, S.S., Gupta, R.S.: Finite Element Vibration Analysis of Rotating Timoshenko Beams. Journal of Sound and Vibration 242, 103–124 (2001)
7. Takahashi, I.: Vibration and Stability of Non-Uniform Cracked Timoshenko Beam Subjected to Follower Force. Computers and Structures 71, 585–591 (1999)
8. Hou, J., Wicks, B.J., Antoniou, R.A.: An Investigation of Fatigue Failures of Turbine Blades in a Gas Turbine Engine by Mechanical Analysis. Engineering Failure Analysis 9, 201–211 (2000)
9. Fisher, R.: The Use of Multiple Measurements in Taxonomic Problems. Ann. Eugenics 7, 178–188 (1936)
10. Nechval, N.A., Purgailis, M., Skiltere, D., Nechval, K.N.: Pattern Recognition Based on Comparison of Fisher's Maximum Separations. In: Proceedings of the 7th International Conference on Neural Networks and Artificial Intelligence (ICNNAI 2012), Minsk, Belarus, pp. 65–69 (2012)
11. Choi, S.W., Park, J.H., Lee, I.B.: Process Monitoring Using a Gaussian Mixture Model via Principal Component Analysis and Discriminant Analysis. Computers and Chemical Engineering 28, 1377–1387 (2004)
12. Chiang, L.H., Russell, E.L., Braatz, R.D.: Fault Diagnosis in Chemical Processes Using Fisher Discriminant Analysis, Discriminant Partial Least Squares, and Principal Component Analysis. Chemometrics and Intelligent Laboratory Systems 50, 243–252 (2000)

Graph Neural Networks for 3D Bravais Lattices Classification

Aleksy Barcz and Stanisław Jankowski

Institute of Electronic Systems, Warsaw University of Technology, Warsaw, Poland
abarcz@gmail.com, s.jankowski@ise.pw.edu.pl

Abstract. This paper presents the computational capabilities of Graph Neural Networks in application to 3D crystal structures. The Graph Neural Network model is described in detail in terms of encoding and unfolded network. Results of classifying selected 3D Bravais lattices are presented, confirming the ability of the Graph Neural Network Model to distinguish between structurally different lattices, lattices containing a single substitution and lattices containing a differently located substitution.

1 Introduction

In the domain of classification and regression we can distinguish two types of data. The first type is the standard, *vectorial* data. Such data can be represented as vectors of a constant length. Each position in such a vector corresponds to a numerical *feature*. Each sample belonging to such data can be described by a vector of features, ordered consistently among the whole dataset. The second type of data is the *structured* or *graph* data. Each sample belonging to such data can be described by a *graph*, consisting of *nodes* and *edges*. Each node or edge can be described by its *label*, which is a constant-length vector of ordered numerical features. However, the explicit information about each sample is stored not only in the node and edge labels, but also in the *structure* of the graph.

The first attempts to process graph data with neural networks models were the Hopfield networks, mapping from one graph node to adjacent nodes [1]. Subsequently, *encoder-decoder* models were invented, like the RAAM [2] and LRAAM model [3]. A single RAAM/LRAAM model usually consists of two fully-connected three layer neural networks (inputs performing no computation, a hidden layer and an output layer). The first network, called the *encoder* is used to build a lossless compressed representation of graph nodes, while the *decoder* network is used to restore the original graph node representation from the compressed code. Such models can be used to build a compact representation of DPAGs (directed positional acyclic graphs), where the representation of a graph root node is treated as the whole graph representation. Such a representation can then be fed to a standard feed-forward neural network for classification/regression purposes. Later on, the Backpropagation Through Structure (BPTS) algorithm was formulated [4], addressing the *moving target* problem and

V. Golovko and A. Imada (Eds.): ICNNAI 2014, CCIS 440, pp. 76–86, 2014.

introducing the idea of the *unfolded network*. Finally, two models were created to process graph data without creating a lossless encoding: the Graph Machines [5] and the Graph Neural Networks [6]. Both models are designed to learn an encoding of graphs which is sufficient for a graph classification/regression task. Such an approach proved to work well in various domains. Graph Machines were successfully used in many QSAR and QSPR tasks [7] [8], including most recently prediction of fuel compounds properties [9]. Graph Neural Networks were used in XML document mining [10], web page ranking [11], spam detection [12], object localisation in images [13] and in image classification [14].

In the field of material technology, the properties of a material often depend on defects in its crystal structure. Such defects may consist of a vacant site in the crystal lattice or of a *basis* substitution. This article describes how three dimensional crystal structures, such as Bravais lattices, can be processed directly by using Graph Neural Networks. By using a graph oriented model, the spacial location of lattice points can be described. By using node labels corresponding to lattice points, information about the basis can be taken into account.

1.1 Related Works

Neural networks were successfully used to predict solid solubility limits for copper and silver, using Bravais lattices parameters as part of the input [15]. To provide the classifier with vectorial data, Bravais lattices were described using unit cell lengths (a, b, c), axes angles (α, β, γ) and the crystal system (simple, base-centered, face-centered, body-centered). In another work, Self-Organising Maps were used to describe structure-property relationships of crystal structures represented using powder diffraction patterns [16]. Both these approaches are representative of the vectorial modelling method. The underlying graph-structured data had to be represented in a form which could be fed to a feed-forward neural network. In this article a different approach is proposed: modelling graph structures by directly using graphs and graph neural network models, which proved useful in other, previously mentioned, graph-based domains.

2 The Graph Neural Network Model

The Graph Neural Network model is a connectionist model, capable of performing classification and regression on almost all types of graphs[6]. In particular, cyclic and acyclic, positional and non-positional, directed and undirected graphs can be processed by this model. A similar solution was the recursive neural network model [17]. However, it is the GNN that combined several existing techniques with a novel training algorithm to produce a very flexible graph processing model.

2.1 Data

A single GNN model is built for a set of graphs (the dataset). Each graph belonging to the dataset can have a different structure. The whole dataset can be thus represented as a single disconnected graph. Each nth graph node is represented by its label $\boldsymbol{l}_n$ of constant size $|\boldsymbol{l}_n| \geq 1$. Each directed edge $u \Rightarrow n$ (from uth to nth node) is represented by edge label $\boldsymbol{l}_{(n,u)}$ of constant size $|\boldsymbol{l}_{(n,u)}| \geq 0$. The edge label size may differ from the node label size. In the implemented model undirected edges were represented as pairs of directed edges to account for the mutual impact of connected nodes. For each nth graph node an output $\boldsymbol{o}_n$ of constant size can be sought - we say it's a *node-oriented* task. Alternatively, a single output $\boldsymbol{o}_g$ can be sought for the whole graph - in such case we say it's a *graph-oriented* task. In this paper a node-oriented approach was chosen, as a potentially more flexible one.

2.2   Computation Units

The GNN model consists of two computational units: the transition unit $f_{\boldsymbol{w}}$ and the output unit $g_{\boldsymbol{w}}$. The vector of parameters $\boldsymbol{w}$ is distinct for each unit, however, for consistency with previous publications it will be denoted by $\boldsymbol{w}$ for both units. The transition unit is used to build the encoded representation of each node, $\boldsymbol{x}_n$ (the *state* of nth node). The output unit is used to calculate the output $\boldsymbol{o}_n$ for each node. Let's denote by $ne[n]$ the set of neighbors of the nth node, that is such nodes u that are connected to the nth node with a directed edge $u \Rightarrow n$. Let's further denote by $co[n]$ the set of directed edges from $ne[n]$ to the nth node. For this implementation the following transition and output functions were used:

$$\boldsymbol{x}_n = f_{\boldsymbol{w}}(\boldsymbol{l}_n, \, \boldsymbol{l}_{co[n]}, \, \boldsymbol{x}_{ne[n]}) \, , \tag{1}$$

$$\boldsymbol{o}_n = g_{\boldsymbol{w}}(\boldsymbol{x}_n) \, . \tag{2}$$

The GNN model offers two forms of the $f_{\boldsymbol{w}}$ function, one better suited for positional graphs, the other one for non-positional graphs. To make processing non-positional graphs possible, the *non-positional form* of $f_{\boldsymbol{w}}$ was chosen [6], which can be described as a simple sum of $h_{\boldsymbol{w}}$ unit outputs. This yields the final set of equations describing the model:

$$\boldsymbol{x}_n = \sum_{u \in ne[n]} h_{\boldsymbol{w}}(\boldsymbol{l}_n, \, \boldsymbol{l}_{(n,u)}, \, \boldsymbol{x}_u) \, , \tag{3}$$

$$\boldsymbol{o}_n = g_{\boldsymbol{w}}(\boldsymbol{x}_n) \, . \tag{4}$$

The $h_{\boldsymbol{w}}$ and $g_{\boldsymbol{w}}$ units were implemented as fully-connected three layer neural networks. Hidden layers consisted for both networks of *tanh* neurons. The output layer of the $h_{\boldsymbol{w}}$ network consisted of *tanh* neurons, as the state $\boldsymbol{x}_n$ must consist of bounded values only. The output layer of the $g_{\boldsymbol{w}}$ network consisted of neurons with linear activation function. During initialization, weights of both networks were set so that the input to every jth neuron, net_j, was bounded assuming normalised input: $net_j \in (-1, 1)$. The initial input weights corresponding to the state inputs were divided by an additional factor, i.e. the maximum node indegree, to take into consideration the fact, that the state consists of a sum of h_w outputs. All the input data (node and edge labels) was normalised before feeding to the model. The f_w and g_w functions are presented in Fig. 1 altogether with one of the corresponding edges, where the comma-separated list of inputs stands for a vector obtained by stacking all the listed values one after another.

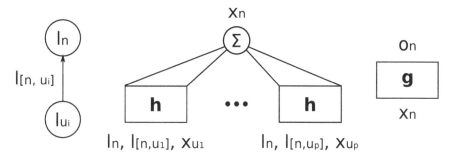

Fig. 1. The f_w and g_w functions for a single node and one of the corresponding edges

2.3 Encoding Network

Graph processing by a GNN model consists of two steps. The first step is to calculate the state x_n for all nodes, using the transition function f_w and exploiting the graph structure. The second step is to process this representation using the output function g_w. The values of node states depend on one another, according to the graph structure, so the whole model can be described in terms of an *encoding network*. The encoding network consists of the f_w unit instances connected according to the graph structure, with an instance of g_w unit connected at each node. It is important to remember, that there exists only one h_w and only one g_w network for the whole GNN. All the instances are just copies of these two networks, using the same sets of weights (the *shared weights* technique).

A sample graph and the corresponding encoding network are presented in Fig. 2. It can be seen that if a graph contains cycles, the state calculation must be an iterative process, to take into account all the node dependencies. Furthermore, we would expect a convergence to be reached, which is also addressed by the GNN model.

2.4 Unfolding and Convergence

To take into account all graph edges (which may form cycles), another representation of the encoding network is used. The encoding network is virtually *unfolded* in time in such a way, that at each time step t_i all f_w units are taken into consideration and the connections between them exist only between subsequent time steps. In such a way all cycles can be processed naturally without

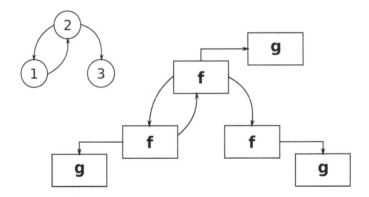

Fig. 2. A sample graph and the corresponding encoding network

additional effort. Let's define the *global state* x as the set of all node states. Let's define l as the set of all node and edge labels. Let's further define o as the set of all node outputs. The *global transition function* F_w and *global output function* G_w, being the unfolded network counterparts of f_w and g_w, are defined as follows:

$$x = F_w(l, x) , \tag{5}$$

$$o = G_w(x) . \tag{6}$$

The global state x is computed at each time step and is expected to converge to $\hat{x}$ after a finite number of steps. Then, the output o_n is calculated by the g_w units. The output error $e_n = (d_n - o_n)^2$ (where d_n stands for the expected output) is calculated and backpropagated through the output units, yielding $\frac{\partial e_w}{\partial o} \cdot \frac{\partial G_w}{\partial x}(\hat{x})$. That value is backpropagated through the unfolded network using the BPTT/BPTS algorithm. Additionally, at each time step the $\frac{\partial e_w}{\partial o} \cdot \frac{\partial G_w}{\partial x}(\hat{x})$ error is injected to the f_w layer. In such a way the error backpropagated through the f_w layer at time t_i comes from two sources. Firstly, it is the original output error of the network $\frac{\partial e_w}{\partial o} \cdot \frac{\partial G_w}{\partial x}(\hat{x})$. Secondly, it is the error backpropagated from the subsequent time layers of the f_w unit from all nodes connected with the given node u by an edge $u \Rightarrow n$. The backpropagation continues until the error value converges, which usually take less time steps than the state convergence. The unfolding and error backpropagation phases are presented in Fig. 3.

The described algorithm makes one important assumption: we want the global state x to converge. To assure convergence, F_w must be a *contraction map*. According to Banach Theorem, this guarantees that the state calculation will converge to a unique point $\hat{x}$ and the convergence will be exponentially fast. To assure the contraction property, a penalty term $\frac{\partial p_w}{\partial w}$ is added to the total $\frac{\partial e_{h_w}}{\partial w}$ error term after performing BPTS. The penalty value p_w is calculated as follows. Let $A = \frac{\partial F_w}{\partial x}(x, l)$ be a block matrix of size $N \times N$ with blocks of size

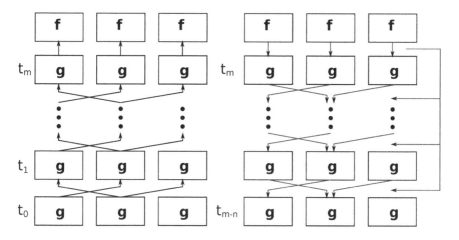

Fig. 3. Unfolded encoding network for the sample graph and backpropagation

$s \times s$, where N is the number of nodes in the processed graph and $|x_n| = s$ is the state size for a single node. A single block $A_{n,u}$ measures the influence of the uth node on the nth node if an edge $u \Rightarrow n$ exists or is zeroed otherwise. Let's denote by I_u^j the influence of uth node on the jth element of state x_n (Eq. 9). The penalty p_w added to the network error e_w is defined by Eq. 7.

$$p_w = \sum_{u \in N} \sum_{j=1}^{s} L(I_u^j, \mu) , \tag{7}$$

$$L(y, \mu) = \begin{cases} y - \mu & \text{if } y > \mu \\ 0 & \text{otherwise} \end{cases} , \tag{8}$$

$$I_u^j = \sum_{(n,u)} \sum_{i=1}^{s} |A_{n,u}^{i,j}| . \tag{9}$$

This does not guarantee that F_w will remain a contraction map, as the penalty is added post factum and it must be tuned using the contraction constant μ. However, even if the convergence isn't always reached, the model can still be trained and yield good results. The necessary constraint in such cases is a maximum number of unfolding and error backpropagation steps.

3 Bravais Lattices

In crystallography, a crystal structure can be described by a *lattice* and a *basis* [18]. The lattice describes the location of lattice points in space, while the basis is a group of atoms (or a single atom) that is placed at each lattice point. A sample lattice, a basis and the resulting two dimensional crystal structure are presented in Fig. 4. A Bravais lattice in three dimensional space is an infinite array of discrete points, generated by translation operations T:

$$T = n_1a + n_2b + n_3c \ . \tag{10}$$

Each translation is described by three constant parameters: a, b and c which are called the *primitive vectors* and three integers: n_1, n_2 and n_3. If we translate the lattice by any such vector T, we will obtain the same lattice. A parallelepiped formed from the primitive vectors is called a *primitive cell*. A primitive cell is a basic component of a crystal structure. We can distinguish 14 Bravais lattices in three dimensions, differing in the primitive vectors length relations, the angles between them and the presence of additional lattice points in the cell.

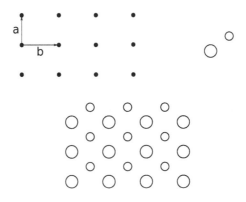

Fig. 4. A sample lattice, a basis and the resulting crystal structure

Many material properties depend on crystal structure defects, which are introduced on purpose. A defect may e.g. consist of an atom missing at one of the lattice points (vacancy) or of a different atom introduced at one of the lattice points (substitution). Properties of materials containing a particular defect are determined experimentally by producing a specimen and testing it in a laboratory. Modelling of such phenomena with neural network models could prove useful for approximating these properties before the actual experiments take place.

4   Experimental Results

For the scope of this work, two simple Bravais lattices were chosen for experiments: the primitive (P) tetragonal lattice and the body-centered (I) tetragonal lattice, containing additional lattice point in the center of each cell. For both tetragonal lattices $a = b \neq c$, as presented in Fig. 5. For all the experiments each crystal structure was represented by a single cell. Each lattice point was described as a graph node, with node label containing information about the basis used at this node. The mutual location of two adjacent points u and n was described as a pair of directed edges, each containing in its label the 3D cartesian coordinates of the vector $u \Rightarrow n$ or $n \Rightarrow u$, respectively. In such way, the description of a cell was independent from the actual location of the cell in space, which was the goal. A spherical coordinates system was also tried out, yielding similar results to the cartesian one.

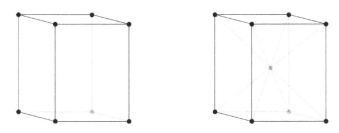

Fig. 5. Simple tetragonal cell (P) and body-centered tetragonal cell (I)

For every experiment the dataset was generated as follows. First a single graph (cell) was created as a cubic (P) cell $a = b = c = 1$. Node labels were set to 1 ($|l_n| = 1$, $l_n = 1$), unless stated otherwise. A second graph, cubic (I) cell, was created by introducing an additional node in the center of the graph. Then, datasets were generated by random scaling of all the input graphs edges using factors pertaining to uniform distribution $U(-5, 5)$, but maintaining the tetragonal constraint: $a = b \neq c$. Then, a small random error ε with zero mean and standard deviation equal to 0.01 was added to all node and edge labels. For every experiment the dataset consisted of two classes of 200 graphs each. Among each class every graph node had its expected output set to the same value: 1 or -1, depending on the class. During evaluation, a node was assigned to class depending on the sign of its output o_n. For every experiment, the training set consisted of 50 graphs, while the test set consisted of 150 graphs.

For every experiment, the state size $|x_n|$ was set to 5. The number of hidden neurons was set for both networks to 5. For the output network linear neurons were used as output neurons. The contraction constant μ was set to 30 as smaller values tended to disturb the learning process significantly. The maximum number of state calculation and error backpropagation steps was set to 200. The number of full training iterations (forward-backward) was set to 200, as it could be seen that for some GNNs the RMSE on training set began to drop significantly only after more than 100 iterations. In each experiment the best GNN was selected as the one that achieved the smallest RMSE on the training set. Then, the selected GNN was evaluated on the test set. The training set for each experiment consisted of 50 samples from each of the two classes (100 samples in total). The test set consisted of 150 samples from each class (300 samples in total).

4.1   Structural Difference

For this experiment, a tetragonal (P) dataset was compared to a tetragonal (I) dataset. All node labels were set to the same value. Thus, the task to solve was to distinguish cells with the central atom missing from full tetragonal (I) cells. The results achieved by the selected GNN are presented in Table 1. It can be seen, that a GNN can be trained to deal very well with such a task, in which the number of nodes in two graphs differs.

Table 1. Structural difference - results

	accuracy	precision	recall
training set	98.94%	98.51%	99.25%
test set	98.82%	98.03%	99.50%

4.2   Single Substitution

For this experiment, two tetragonal (I) datasets were used. In one dataset all node labels were set to 1, while in the other one the labels of central nodes were set to 2, to simulate a single atom substitution. The unusually good results for the small random error, presented in Table 2, can be explained by the simplicity of the task. As node labels are explicitly given to the model (not alike the structure of the graph, which was the case described in the previous section), a simple linear classifier would be sufficient for this task, even taking into consideration the noise applied.

To check the model performance with a more demanding task, a larger random error ε with zero mean and standard deviation equal to 0.1 was added to the original node and edge labels and a new classifier was trained. The results are significantly worse, however, it must be stated, that such a random error disturbs greatly the graph structure in the case of small edge lengths.

Table 2. Single substitution - results

	accuracy	precision	recall
training set ($sd\ \varepsilon = 0.01$)	100%	100%	100%
test set ($sd\ \varepsilon = 0.01$)	100%	100%	100%
training set ($sd\ \varepsilon = 0.1$)	96.00%	96.00%	96.00%
test set ($sd\ \varepsilon = 0.1$)	94.92%	95.29%	94.51%

4.3   Substitution Location Difference

For this experiment, two tetragonal (I) datasets were used, both of them containing a single substitution (node label of one node set to 2). However, for the first set the substituted node was the central node, while for the other set one of the corner nodes was substituted. Therefore, every graph had eight nodes labelled as 1 and a single node labelled as 2. The location of the substituted node was different for the two sets and the task was to find this difference basing on nodes connections. This task proved to be the most difficult for the GNN model. Not only the difference in node labels was to be taken into consideration, but also the different placement of the substituted node, so the structure of the graph had to be exploited properly. Nevertheless, the GNN model proved to achieve very good results in this task, as presented in Table 3.

Table 3. Substitution location difference - results

	accuracy	precision	recall
training set	96.33%	94.26%	98.66%
test set	95.07%	92.05%	98.66%

5   Conclusion

The Graph Neural Network is a model successfully used in many two dimensional graph processing tasks. This article presents the capabilities of the GNN model to process three dimensional data, such as crystal structures. The main difference of this data lays in the fact, that the spatial location of all the nodes must be taken into consideration and not only the edge and node properties. The tasks used for testing the GNN model were selected so as to present how the GNN can be used to deal with various structural and property differences. The model proved to work well in all the tasks, including a simulated crystal structure vacancy defect and two different atom substitution defects.

References

1. Goulon-Sigwalt-Abram, A., Duprat, A., Dreyfus, G.: From hopfield nets to recursive networks to graph machines: numerical machine learning for structured data. Theoretical Computer Science 344(2), 298–334 (2005)
2. Pollack, J.B.: Recursive distributed representations. Artificial Intelligence 46(1), 77–105 (1990)
3. Sperduti, A.: Labelling recursive auto-associative memory. Connection Science 6(4), 429–459 (1994)
4. Goller, C., Kuchler, A.: Learning task-dependent distributed representations by backpropagation through structure. In: IEEE International Conference on Neural Networks, vol. 1, pp. 347–352. IEEE (1996)

5. Goulon, A., Duprat, A., Dreyfus, G.: Learning numbers from graphs. In: Applied Statistical Modelling and Data Analysis, Brest, France, pp. 17–20 (2005)
6. Scarselli, F., Gori, M., Tsoi, A.C., Hagenbuchner, M., Monfardini, G.: The graph neural network model. IEEE Transactions on Neural Networks 20(1), 61–80 (2009)
7. Goulon, A., Picot, T., Duprat, A., Dreyfus, G.: Predicting activities without computing descriptors: graph machines for QSAR. SAR and QSAR in Environmental Research 18(1-2), 141–153 (2007)
8. Goulon, A., Faraj, A., Pirngruber, G., Jacquin, M., Porcheron, F., Leflaive, P., Martin, P., Baron, G., Denayer, J.: Novel graph machine based QSAR approach for the prediction of the adsorption enthalpies of alkanes on zeolites. Catalysis Today 159(1), 74–83 (2011)
9. Saldana, D., Starck, L., Mougin, P., Rousseau, B., Creton, B.: On the rational formulation of alternative fuels: melting point and net heat of combustion predictions for fuel compounds using machine learning methods. SAR and QSAR in Environmental Research 24(4), 259–277 (2013)
10. Yong, S., Hagenbuchner, M., Tsoi, A., Scarselli, F., Gori, M.: XML document mining using graph neural network. Center for Computer Science, 354 (2006), http://inex.is.informatik.uni-duisburg.de/2006
11. Scarselli, F., Yong, S.L., Gori, M., Hagenbuchner, M., Tsoi, A.C., Maggini, M.: Graph neural networks for ranking web pages. In: Proceedings of the 2005 IEEE/WIC/ACM International Conference on Web Intelligence, pp. 666–672. IEEE (2005)
12. Scarselli, F., Tsoi, A.C., Hagenbuchner, M., Noi, L.D.: Solving graph data issues using a layered architecture approach with applications to web spam detection. Neural Networks 48, 78–90 (2013)
13. Monfardini, G., Di Massa, V., Scarselli, F., Gori, M.: Graph neural networks for object localization. Frontiers in Artificial Intelligence and Applications 141, 665 (2006)
14. Quek, A., Wang, Z., Zhang, J., Feng, D.: Structural image classification with graph neural networks. In: International Conference on Digital Image Computing Techniques and Applications (DICTA), pp. 416–421. IEEE (2011)
15. Zhang, Y., Yang, S., Evans, J.R.G.: Revisiting Hume-Rotherys Rules with artificial neural networks. Acta Materialia 56(5), 1094–1105 (2008)
16. Willighagen, E., Wehrens, R., Melssen, W., De Gelder, R., Buydens, L.: Supervised self-organizing maps in crystal property and structure prediction. Crystal Growth & Design 7(9), 1738–1745 (2007)
17. Bianchini, M., Maggini, M., Sarti, L., Scarselli, F.: Recursive neural networks for processing graphs with labelled edges: Theory and applications. Neural Networks 18(8), 1040–1050 (2005)
18. Kittel, C., McEuen, P.: Introduction to solid state physics, vol. 8. Wiley, New York (1986)

Quality Evaluation of E-commerce Sites
Based on Adaptive Neural Fuzzy Inference System

Huan Liu[1,2] and Viktor V. Krasnoproshin[2]

[1] School of Software, Harbin University of Science and Technology, No. 994 Mailbox,
23 Sandadongli Road, Xiangfang District, Harbin, 150040, China
hliu@hrbust.edu.cn
[2] Faculty of Applied Mathematics and Computer Science, Belarusian State University,
4 Nezavisimost av., Belarusian State University, Minsk, 220030, Belarus
krasnoproshin@bsu.by

Abstract. This paper describes a combined approach to the intelligent evaluation problem of E-commerce sites. The methodology of adaptive neural networks with fuzzy inference was used. A model of a neural network was proposed, in the frame of which expert fuzzy reasoning and rigorous mathematical methods were jointly used. The intelligent system with fuzzy inference was realized based on the model in Matlab software environment. It shows that the system is an effective tool for the quality analysis process modelling of the given type of sites. It also shows that the convenient and powerful tool is much better than the traditional artificial neural network for the simulation of sites evaluation.

Keywords: Neural network, quality evaluation, Adaptive Neural Fuzzy Inference System (ANFIS), fuzzy logic, E-commerce website.

1   Introduction

In the rapid development of global networks gained wide popularity with businesses of electronic commerce (also called "e-commerce" or "eCommerce"). This term covers a wide range of activities of modern enterprises. It includes the entire Internet - the process for the development, marketing, sale, delivery, maintenance, payment for goods and services.

The key to e-commerce companies are problems of understanding customer inquiries and the development of tools for the implementation of feedback. Companies is usually presented poorly online with sites, which are difficult for the reaction. This significantly weakens the position of the company as a whole. Consequently, it is very important that businesses have the opportunity to assess the quality of their business proposals and understand how customers perceive them in the context of the industry [1,2].

Therefore, it plays an important role that e-commerce companies assess their sites for successful operation. Ratings are a kind of feedback mechanism that allows refining the strategy and methods of control. Website is a software product that can be

V. Golovko and A. Imada (Eds.): ICNNAI 2014, CCIS 440, pp. 87–97, 2014.
© Springer International Publishing Switzerland 2014

considered as a system with a sufficiently complex structure and function. Its importance as a basic element of e-commerce is impossible to assess from the perspective of only one criterion. Therefore, the problem of assessing sites refers to the classification of multi-criteria optimization, which takes source data typically used subjective information from experts [2,3].

At present, two main approaches to solving the problem of evaluation of e-commerce sites: quantitative when building a numerical score, and qualitative when the resulting estimate is described by some kind of linguistic expression "as good (or bad)". Because of the weak formalization problem, all algorithms developed in the framework of these approaches are heuristic. They are based on knowledge and experience of the researcher and are a set of systematic steps in some way without the relationship of factors affecting the final decision [4,5]. In this paper, we are focusing on the assessing problem, for solving which we built an adaptive evaluation system to assess the E-commerce website, on the base of previous assessments projects from experts.

Therefore, we propose a combined approach for solving the problem of estimating the quality of e-commerce sites. In this approach, expert information and rigorous mathematical methods are effectively used. As a basic tool offers the possibility to use an adaptive neural fuzzy inference system (ANFIS), which belongs to a class of hybrid intelligent systems. A model of adaptive neural network was implemented intelligent fuzzy inference in the Matlab programming environment. It shows that the system is a convenient and powerful tool for the simulation of assessment sites, and it uses the "if-then" type rules that are easy to understand and interpret.

2   Modeling Based on Adaptive Neural Network Fuzzy Inference

2.1   Description of the Problem

For the above purpose, at first, we constructed the knowledge base from these projects; then we designed and realized the evaluation ANFIS; at last, we trained and tested the ANFIS. Each of previous E-commerce website assessment projects can be expressed as follows:

$$P =< S, FEM >,$$
$$S = \{EM^{C_i}\}, \quad for \ i = 1...n \tag{1}$$

where S is the set of evaluation marks of all criteria, FEM is the final evaluation mark, EM^{C_i} is evaluation mark i_{th} criterion C_i, and n is the total amount of criteria. Moreover, the EWAS can be described as:

$$O =< PS^{Tr}, PS^{Ts}, S, R >,$$
$$PS^{Tr} = \{P^i\},$$
$$PS^{Ts} = \{P^{m-i}\}, \quad for \ i = 1...m \tag{2}$$

where PS^{Tr} is training set from part of all projects, PS^{Ts} is testing set from the rest part of all projects, S is set of evaluation marks of all criteria, and R is final evaluation mark from our ANFIS.

2.2 ANFIS Method

Analysis of existing literature [6,7,8,9] shows that ANFIS has good opportunities for learning, prediction and classification. The architecture of these networks allows adaptively based on numerical or expertise data to create a knowledge base (in the form of a set of fuzzy rules) for the system output.

ANFIS is a multilayer unidirectional neural learning network, which uses fuzzy reasoning. Figure 1. shows a typical ANFIS architecture with two entrances, four rules and one output. Each input network mapped two membership functions (MF).

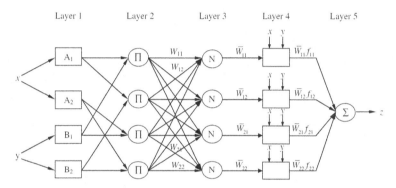

Fig. 1. ANFIS architecture with two inputs and four rules

In this model, the first-order rules are used if-then type that are given by:

Rule 1: *If x is* A_1 *and y is* B_1 *then* $f_{11} = p_{11}x + q_{11}y + r_{11}$
Rule 2: *If x is* A_1 *and y is* B_2 *then* $f_{12} = p_{12}x + q_{12}y + r_{12}$
Rule 3: *If x is* A_2 *and y is* B_1 *then* $f_{11} = p_{21}x + q_{21}y + r_{21}$
Rule 4: *If x is* A_2 *and y is* B_2 *then* $f_{12} = p_{22}x + q_{22}y + r_{22}$

A_1, A_2, B_1 and B_2 are membership functions of inputs x and y, respectively, p_{ij}, q_{ij} and r_{ij} ($i, j = 1,2$) are the output parameters.

Reduced network architecture consists of five layers. In order to continue to use this type of architecture, each layer are described in details.

Layer 1: The input nodes. All nodes are adaptive layer. They generate membership grades to which they belongs to each of the appropriate fuzzy sets by the following formulas:

$$O_{A_i}^1 = \mu_{A_i}(x) \quad i = 1, 2,$$

$$O_{B_j}^1 = \mu_{B_j}(y) \quad j = 1, 2, \tag{3}$$

where x and y where x and y are crisp inputs, and A_i and B_j are fuzzy sets such as low, medium, high characterized by appropriate MFs, which could be triangular, trapezoidal, Gaussian functions or other shapes. In this study, the generalized bell-shaped MFs defined below are utilized

$$\mu_{A_i}(x) = \frac{1}{1 + \left(\dfrac{x - c_i}{a_i}\right)^{2b_i}}, \quad i = 1, 2,$$

$$\mu_{B_j}(y) = \frac{1}{1 + \left(\dfrac{y - c_j}{a_j}\right)^{2b_j}}, \quad j = 1, 2, \tag{4}$$

where $\{a_i, b_i, c_i\}$ and $\{a_j, b_j, c_j\}$ are the parameters of the MFs, governing the bell-shaped functions. Parameters in this layer are referred to as premise parameters.

Layer 2: The nodes in this layer are fixed nodes labelled, indicating that they perform as a simple multiplier. The outputs of this layer are represented as

$$O_{ij}^2 = W_{ij} = \mu_{A_i}(x)\mu_{B_j}(y), \quad i, j = 1, 2, \tag{5}$$

which represents the firing strength of each rule. The firing strength means the degree to which the antecedent part of the rule is satisfied.

Layer 3: The nodes in this layer are also fixed nodes labelled N, indicating that they play a normalization role in the network. The outputs of this layer can be represented as

$$O_{ij}^3 = \overline{W}_{ij} = \frac{W_{ij}}{W_{11} + W_{12} + W_{21} + W_{22}}, \quad i, j = 1, 2, \tag{6}$$

which are called normalized firing strengths.

Layer 4: Each node in this layer is an adaptive node, whose output is simply the product of the normalized firing strength and a first-order polynomial (for a first-order Sugeno model). Thus, the outputs of this layer are given by

$$O_{ij}^4 = \overline{W}_{ij} f_{ij} = \overline{W}_{ij} \left(p_{ij} x + q_{ij} y + r_{ij}\right), \quad i, j = 1, 2, \tag{7}$$

Parameters in this layer are referred to as consequent parameters.

Layer 5: The single node in this layer is a fixed node labelled Σ, which computes the overall output as the summation of all incoming signals, i.e.

Parameters in this layer are referred to as consequent parameters.

$$z = O_1^5 = \sum_{i=1}^{2}\sum_{j=1}^{2}\overline{W}_{ij} f_{ij} = \sum_{i=1}^{2}\sum_{j=1}^{2}\overline{W}_{ij}\left(p_{ij}x + q_{ij}y + r_{ij}\right)$$

$$= \sum_{i=1}^{2}\sum_{j=1}^{2}\left[(\overline{W}_{ij}x)p_{ij} + (\overline{W}_{ij}y)q_{ij} + (\overline{W}_{ij})r_{ij}\right]$$

(8)

which is a linear combination of the consequent parameters when the values of the premise parameters are fixed.

It can be observed that the ANFIS architecture has two adaptive layers: Layers 1 and 4. Layer 1 has modifiable parameters $\{a_i, b_i, c_i\}$ and $\{a_j, b_j, c_j\}$ related to the input MFs. Layer 4 has modifiable parameters $\{p_{ij}, q_{ij}, r_{ij}\}$ pertaining to the first-order polynomial. The task of the learning algorithm for this ANFIS architecture is to tune all the modifiable parameters to make the ANFIS output match the training data. Learning or adjusting these modifiable parameters is a two-step process, which is known as the hybrid learning algorithm. In the forward pass of the hybrid learning algorithm, the premise parameters are hold fixed, node outputs go forward until Layer 4 and the consequent parameters are identified by the least squares method. In the backward pass, the consequent parameters are held fixed, the error signal propagate backward and the premise parameters are updated by the gradient descent method. The detailed algorithm and mathematical background of the hybrid-learning algorithm can be found in [9].

3   E-commerce Site Evaluation Using AHCHB

This section presents the development of an ANFIS for E-commerce websites assessment and tests its performance.

3.1   Data Description

The dataset used for developing an ANFIS was provided by the Heilongjiang Institute of Information. The dataset contains 507 E-commerce website assessment (EWA) projects and is randomly split into two sample sets: training dataset with 390 projects and testing dataset with 117 projects. Both the training and testing cover all levels and types of E-commerce website assessment.

Inputs to the ANFIS are the usability (U), reliability (R) and design (d) from the 507 E-commerce website assessment projects, which all range from 0 to 4 with 0 representing very bad, 1 bad, 2 normal, 3 good and 4 very good. Output to the ANFIS is the assessment scores (ASs) of the 507 assessment projects, which range from 5 to 99, as shown in Fig. 2.

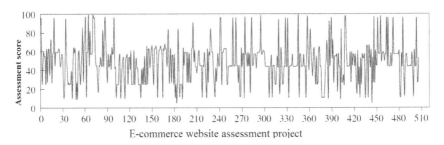

Fig. 2. E-commerce website assessment scores of 506 assessment projects

3.2   Development of the ANFIS

With the training dataset, we choose two generalized bell-shaped MFs for each of the three inputs to build the ANFIS, which leads to 27 if–then rules containing 104 parameters to be learned. Note that it is inappropriate to choose four or more MFs for each input; because the parameters needing to be learned in that case will be greater than the number of training samples. Fig. 3 shows the structure of the ANFIS that is to be built for E-commerce website assessment in this study. The model structure is implemented using the fuzzy logic toolbox of MATLAB software package [10].

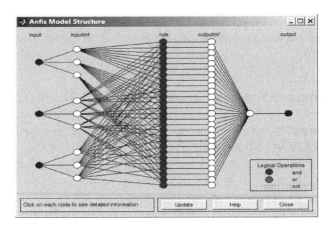

Fig. 3. Model structure of the ANFIS for E-commerce website assessment

The trained if–then rules are presented in Fig. 4, which can be used for prediction. For example, if we change the values of the four inputs from 1.5 to 3, then we immediately get the new output value of the ANFIS as 75.7. This is illustrated in Fig. 5.

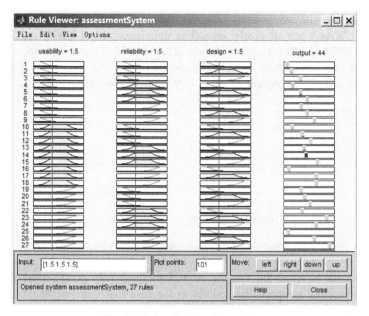

Fig. 4. "If–then" rules after training

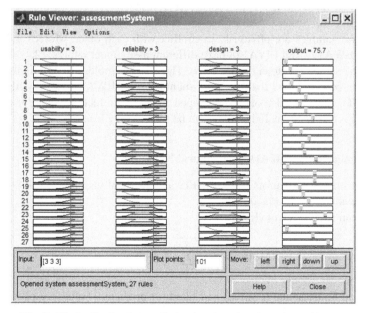

Fig. 5. "If–then" rules for prediction by changing the values of inputs

The trained ANFIS is validated by the testing dataset. Fig. 6 shows the testing errors for the testing dataset. For convenience, the fitting errors for the training dataset are also shown in Fig. 8, from which it can be observed that except for three

E-commerce website assessments (EWAs), the fitting and testing errors for all the other 504 EWAs s are all nearly zero. The three exceptional EWAs are EWA178, EWA407 and EWA446, whose assessment scores are 56, 77 and 77, respectively, but the fitted or predicted values for them by the ANFIS are 44, 83 and 83. The relative errors for these three EWAs are 21.43%, 7.79% and 7.79%, respectively.

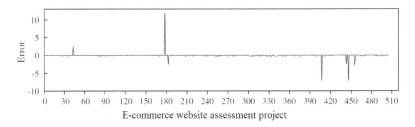

Fig. 6. Fitting and testing errors of the assessment projects by ANFIS

It should not be expected that the ANFIS produce very good results for all the training and testing samples. Particularly in the case that there might be conflicting data in the training or testing dataset.

Looking into the training dataset, we find that both EWA170 and EWA178 have the same assessment ratings: 2, 2, 3 for U, R, D, respectively, but different assessment scores: 44 and 56.

These two samples are obviously in conflict with each other. Moreover, it is also found that EWA177 and EWA178 have different assessment ratings: (2, 3, 1) and (2, 1, 1), but the same assessment score of 56. These two samples also conflict with each other. It can be concluded that the assessment score of EWA178 is very likely to be an outlier. The performance of the developed ANFIS is in fact very good if the fitting error of EWA178 is not included. This can be seen clearly from Fig. 8.

3.3   Comparisons with Artificial Neural Network

To compare the performances of the ANFIS and artificial neural network (ANN), the following evaluation criteria are adopted.

Root mean squared error (RMSE):

$$\text{RMSE} = \sqrt{\frac{1}{N}\sum_{t=1}^{N}(A_t - F_t)^2}, \tag{9}$$

where A_t and F_t are actual (desired) and fitted (or predicted) values, respectively, and N is the number of training or testing samples.

Mean absolute percentage error (MAPE):

$$\text{MAPE} = \frac{1}{N}\sum_{t=1}^{N}\left|\frac{A_t - F_t}{A_t}\right| \times 100 \tag{10}$$

Correlation coefficient (R):

$$R = \frac{\sum_{t=1}^{N}(A_t - \overline{A})(F_t - \overline{F})}{\sqrt{\sum_{t=1}^{n}(A_t - \overline{A})^2 \cdot \sum_{t=1}^{n}(F_t - \overline{F})^2}} \tag{11}$$

where $\overline{A} = \dfrac{1}{N}\sum_{t=1}^{N} A_t$ and $\overline{F} = \dfrac{1}{N}\sum_{t=1}^{N} F_t$ are the average values of A_t and F_t over the training or testing dataset. The smaller RMSE and MAPE, larger R means better performance.

According to [11], the best ANN structure for the 507 E-commerce website assessment projects is a three layer back propagation network with 10 hidden neurons, as shown in Fig. 7. The performances of the ANFIS and ANN in modelling E-commerce website assessment are presented in Table 1, where the two models are trained using the same training dataset and validated by the same testing dataset.

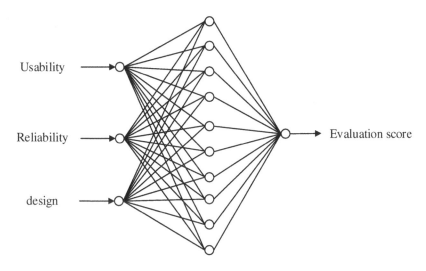

Fig. 7. ANN architecture for E-commerce website assessment

Fig. 8 shows the fitting and testing errors for the 507 E-commerce website assessment projects obtained by the ANN. It is very clear from Table 1 and Figs. 6 and 8 that the ANFIS has smaller RMSE and MAPE as well as bigger R for both the training and testing datasets than the ANN model. In other words, the ANFIS achieves better performances than the ANN model. Therefore, ANFIS is a good choice for modelling E-commerce website assessment.

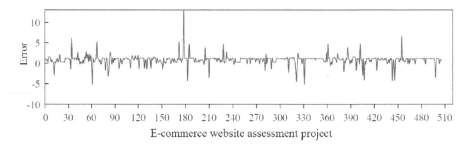

Fig. 8. Fitting and testing errors of the assessment projects by ANN

Table 1. Performances of ANFIS and ANN in modelling E-commerce website assessment

Model	Training dataset			Testing dataset		
	RMSE	MAPE(%)	R	RMSE	MAPE(%)	R
ANFIS	0.57	0.36	0.9994	0.98	0.59	0.9992
ANN	1.64	3.19	0.9963	1.58	3.63	0.9961

Moreover, ANN is a black box in nature and its relationships between inputs and outputs are not easy to be interpreted, while ANFIS is transparent and its if–then rules are very easy to understand and interpret. However, the drawback of ANFIS is its limitation to the number of outputs. It can only model a single output.

4   Conclusion

Evaluation of E-commerce sites is an urgent problem for the majority of e-business enterprises. Within the effective use expert information and rigorous mathematical methods, "if-then" type rules are easily understand and interpreted. An adaptive neural fuzzy inference system was described. After building, training and testing, the proposed ANFIS can evaluate E-commerce site by experts' expressions. It is shows that the convenient and powerful tool is much better than the ANN for the simulation of sites evaluation.

Thus, the use of mathematical models and methods significantly reduces the amount of resources and time, and are necessary to obtain sites assessment results in unconventional decision-making.

Acknowledgements. This research was supported by "Science and Technology Research Project of the Education Department of Heilongjiang Province" in 2014, under the project "Intelligent multi-criteria decision-making method and applied research based on fuzzy AHP and neural network" with project No.12541150.

References

1. Lee, S.: The effects of usability and web design attributes on user preference for e-commerce web sites. Computers in Industry 61(4), 329–341 (2010)
2. Liu, H., Krasnoproshin, V., Zhang, S.: Fuzzy analytic hierarchy process approach for E-Commerce websites evaluation. World Scientific Proceedings Series on Computer Engineering and Information Science 6, 276–285 (2012)
3. Liu, H., Krasnoproshin, V., Zhang, S.: Algorithms for Evaluation and Selection E-Commerce Web-sites. Journal of Computational Optimization in Economics and Finance 4(2-3), 135–148 (2012)
4. Liu, H., Krasnoproshin, V., Zhang, S.: Combined Method for E-Commerce Website Evaluation Based on Fuzzy Neural Network. Applied Mechanics and Materials 380-384, 2135–2138 (2013)
5. Law, R., Qi, S., Buhalis, D.: Progress in tourism management: A review of website evaluation in tourism research. Tourism Management 31(3), 297–313 (2010)
6. Hung, W., McQueen, R.J.: Developing an evaluation instrument for e-commerce web sites from the first-time buyer's viewpoint. Electron. J. Inform. Syst. Eval. 7(1), 31–42 (2004)
7. Azamathulla, H.M., Ghani, A.A., Fei, S.Y., Azamathulla, H.M.: ANFIS-based approach for predicting sediment transport in clean sewer. Applied Soft Computing 12(3), 1227–1230 (2012)
8. Dwivedi, A.A., Niranjan, M., Sahu, K.: Business Intelligence Technique for Forecasting the Automobile Sales using Adaptive Intelligent Systems (ANFIS and ANN). International Journal of Computer Applications 74(9), 7–13 (2013)
9. Jang, J.S.R.: ANFIS: Adaptive-network-based fuzzy inference systems. IEEE Transactions on Systems Man and Cybernetics 23, 665–685 (1993)
10. Petković, D., Issa, M., Pavlović, N.D.: Adaptive neuro-fuzzy estimation of conductive silicone rubber mechanical properties. Expert Systems with Applications 39(10), 9477–9482 (2012)
11. Tsai, C.F., Wu, J.W.: Using neural network ensembles for bankruptcy prediction and credit scoring. Expert Systems with Applications 34(4), 2639–2649 (2008)
12. Singh, R., Kainthola, A., Singh, T.N.: Estimation of elastic constant of rocks using an ANFIS approach. Applied Soft Computing 12(1), 40–45 (2012)

Visual Fuzzy Control for Blimp Robot
to Follow 3D Aerial Object

Rami Al-Jarrah and Hubert Roth

Department of Automatic Control Engineering
Siegen University
Hoelderlinstr. 3, Siegen 57068 Germany
{rami.al-jarrah,hubert.roth}@uni-siegen.de

Abstract. This works presents a novel visual servoing system in order to follow a 3D aerial moving object by blimp robot and to estimate the metric distances between both of them. To realize the autonomous aerial target following, an efficient vision-based object detection and localization algorithm is proposed by using Speeded Up Robust Features technique and Inverse Perspective Mapping which allows the blimp robot to obtain a bird's eye view. The fuzzy control system is relies on the visual information given by the computer vision algorithm. The fuzzy sets model were introduced imperially based on possibilities distributions and frequency analysis of the empirical data. The system is focused on continuously following the aerial target and maintaining it within a fixed safe distance. The algorithm showing robustness against illumination changes , rotation invariance as well as size invariance. The results indicate that the proposed algorithm is suitable for complex control missions.

1 Introduction

Recently, the developing in computer vision makes it very important part on the robot researches. In order to provide the UAVs with additional information to perform visually guided missions, the developing of computer vision techniques are necessary. The robot vision tracking technology is utilized widely due to its advantages such as reliability and low cost. It is applied widely in the fields of security, traffic monitoring, and object recognition. Actually, there are several researches have been presented and devoted to vision flight control, navigation, tracking and object identification [1,2,3]. In [4], they have proposed a visual information algorithm for cooperative robotics. In addition, the visual information has been proposed on aerial robotics for flying information [5]. Another approach also has been proposed for fixed wind UAV flying at constant altitude following circular paths in order to pursuit a moving object on a ground planar surface [6]. In addition, the autonomous surveillance blimp system is provided with one camera as a sensor. The motion segmentation to improve the detection and tracking of point features on a target object was presented in [7] . However, it has some disadvantages like sensitivity to light and the amount of image data. Some studies have proposed many vision tracking methods

V. Golovko and A. Imada (Eds.): ICNNAI 2014, CCIS 440, pp. 98–111, 2014.
© Springer International Publishing Switzerland 2014

such as the simple color tracking [8], template matching [9], background subtraction [10], feature based approaches [11] and feature tracking [12].

However, many algorithms could not meet the requirements of real time and robustness due to large amount of video data. Therefore, it becomes important to improve the accuracy and real time of tracking algorithms. In order to improve the detection of targets in real time and robustness, Lowe has proposed the Scale Invariant Features Transform (SIFT) [13], and then the Speeded Up Robust Features (SURF) has been proposed [14]. Because the processing time of SURF algorithm is faster than that of SIFT, interest point detection of SURF algorithm is used for real-time processing. The visual servoing has been implemented on UAVs successfully. Pose-base methods have been employed for some applications like autonomous landing on moving objects. These methods are important in order to estimate the 3D object position [15]. The image based methods also have been used widely for positioning [16]. In [17] they deal with this problem by proposed a measurement model of the vision sensor based on the specific image processing technique that is related to the size of the target. However, a nonlinear adaptive observer is implemented for the problem and the performance of the proposed method was verified through numerical simulations. For the bearing measurement sensor, the maximization of determinant and fisher information matrix have been studied to generate the optimal trajectory [18,19].

Moreover, these approaches are hard to implement because of the high computational load. In addition, the tracking problem has been studied by using a robust adaptive observer and the intelligent excitation concept [20-22]. Note that the observer is only applied to the systems whose relative degree is 1, and therefore it could not be implemented to higher relative degree systems. The 3D object following method has been developed based on visual information to generate a dynamic look and move control architecture for UAV [23]. Also, aerial object following using visual fuzzy servoing has been presented [24]. However, they approach the problem of the tracking by exploiting the color characteristic of the target which means define a basic color to the target and assuming a simple colored mark to track it. This process is not always perfect and might face problems due to the changes in color over time. Also, the design of the fuzzy controllers were made based on the excellent results after trials and errors method. Actually, the intelligent controllers are getting more importance in robotics researches and one of the most techniques used in intelligent computing techniques is the fuzzy logic.

The most important problem in the fuzzy logic is how to design the fuzzy knowledge base without relies on a human control expert or simulation studies. Recently, the possibility theory which is an alternative to probability theory becomes very important in robot research. It is a mathematical notion that deals with certain types of uncertainties in any system. The theory itself is not only strongly linked to fuzzy systems in its mathematics, but also in its semantics [25]. The possibility theory was introduced by Lotfi Zadeh [26] as an extension of his theory of fuzzy sets and fuzzy logic and then developed by Dubois et al [27]. Therefore, the combination between the possibility theory and fuzzy sets leads to model the complex systems empirically without regard to the presence of the expert. As the possibilities theory deals only with evidence in where the focal elements are overlapping, it is always better to collect possibilities data empirically in Labs. Joslyn Cliff [28] used interval statistics sets with their empirical random sets to develop an approach to construct the possibility distribution histogram.

In this paper, we present the design and implementation of vision system for a blimp robot. The proposed vision is designed by using SURF to localize the 3D aerial target. In addition, the Inverse Perspective Mapping (IPM) has been used to allow to remove the perspective effect from the image and to remap the image into a new 2-D domain where the information content is homogeneously distributed among all pixels. This approach will not depend on the colour, shape or even the size of the tracked object. Moreover, it depends on the target itself by finding the interest points of the objects relies on the SURF and IPM algorithms. Hence, the proposed approach can estimate the metric distances between the blimp robot and the 3D flying object. Then, a fuzzy sets model will be designed imperially to advanced and efficient design the fuzzy controllers to command the blimp to track , follow the aerial target, and keep it maintain with a fixed distance. The blimp system would achieve to be robust object recognition even when the object in a captured image is at a scale in-variance and/or at rotation in-variance compared to the reference images.

The reminder of this paper is organized as follows: after the introduction section we will discuss the blimp hardware of the vision system. Section 3 presents the vision-based robot localization. Then, we introduce the fuzzy sets model based on possibilities histograms in section 4. The experimental results of the vision system obtained through actual flight tests are presented in Section 5. Finally, section 6 presents the conclusion of the work.

2   Blimp Hardware

In order to develop a blimp based on a small size, light weight, high level functionality and communications, the navigation system and autonomous embedded blimp system were presented in our previous works [29, 30]. The core of the blimp system is distributed among the Gumstix Overo-Air COM (weight 5.6g) and Atmega328 microcontroller. The Gumstix will process the data of the sensors and then, execute the commands to the Atmega via I2C protocol . Since the weight is especially important as the mass budget will be one of the most restricting criteria during design, a Gumstix Caspa Camera was chosen due to its weight 22.9 grams.

3   Vision Based Object Localization

A visual system has been used in order to localize the flying target. We processed the captured images from the camera to localize the flying target inside the image. Then, we made a correspondence between the location on the image and the location on work field. Because we seek a detection method which allows having good detection with scaled invariant, rotation invariant and robust against noise, and several approaches had been studied and the examination shows that SURF is the most convenient for our demands. We proposed the SURF algorithm in our previous work [31] in order to detect and track a ground robot. However, in this work we extended and updated the algorithm by using IPM algorithm. This combination gives the blimp ability not only to obtain the bird's eye view, but also to estimate the distances between the blimp and the 3D object. In addition, this work deals with detecting and tracking flying object in the space. As the SURF needs a robust estimator to filter the total set of matched points and eliminate erroneous matches, the Random Sample Consensus RANSAC algorithm is used to discard the outliers from the set of matched points [32]. Considering a flying object at a certain altitude H_o on the world space. The blimp robot has a certain known altitude H_b with position P_b and the flying object location is P_o, The distance between both of them is D in ground plane as shown in Fig 1.

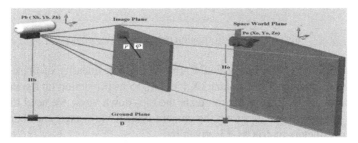

Fig. 1. Coordinate of the Flying object related to Camera

It was assumed that the position of the blimp robot is known when the mission starts. Therefore, These data might be used in order to localize the 3D object with respect to the blimp position as it is given by:

$$P_o = (X_b - H_o, Y_b, Z_b + D) \tag{1}$$

by estimating the distance between both of them and keep the 3D object at the center of the image, Eq. 1 could be valid to find the object position. Then, 3D object projection on the image plane defines by its center point in (x_i, y_i) plane with polar coordinate (r, ϕ) where:

$$r = \sqrt{x_i^2 + y_i^2}, \Phi = \tan^{-1}(x_i / y_i) \tag{2}$$

where r is the distance between the projection point of the abject and the projection point of the camera onto the image plane. ϕ is the angle between object and center line of image plane. In fact, the perspective view of captured image somehow distorts the actual shape of the space in (X_w, Y_w, Z_w) world coordinate. The angle of view and the distances of the object from the camera contribute to associate a different information to each pixel of the image .Hence, this image needs to go through a pre-processing step to remedy this distortion by using the transformation technique known as the Inverse Perspective Mapping (IPM) [33]. The IPM allows to remove the perspective effect from the image and to remap the image into a new 2-D domain where the information content is homogeneously distributed among all pixels. If we assume that the plane of 3D flying object is planner as it is shown in Fig 2. The rotation value which is the translation along the camera optical axis is θ. However, the use of single camera ,which is mounted on the gondola, does not provide the depth information about this object because the non linearity between the object position in the image plane and its position in the 3D object plane as well as in the real world plane.

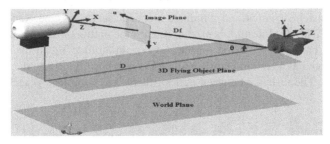

Fig. 2. IPM transformation between the two objects

The transforming of the image could remove the non linearity of these distances. The mapping of the object in its plane (X_o, Y_o, Z_o) to its projection on the image plane (u, v) is shown in Fig. 2. In order to create the top- down view, we need to rotate the image with angle θ, translate along the camera optical view axis, and then, scale by the camera parameters matrix as it is given by:

$$[u, v, 1]^T = KTR \, (x, y, z, 1)^T \tag{3}$$

$$R = \begin{bmatrix} 1 & 0 & 0 & 0 \\ 0 & \cos\theta & -\sin\theta & 0 \\ 0 & \sin\theta & \cos\theta & 0 \\ 0 & 0 & 0 & 1 \end{bmatrix}, T = \begin{bmatrix} 1 & 0 & 0 & 0 \\ 0 & 1 & 0 & 0 \\ 0 & 0 & 1 & -h/\sin\theta \\ 0 & 0 & 0 & 1 \end{bmatrix}, K = \begin{bmatrix} \alpha_x & s & u_o & 0 \\ 0 & \alpha_y & v_o & 0 \\ 0 & 0 & 1 & 0 \end{bmatrix} \tag{4}$$

where R, T, K are the rotation, translation, and camera parameters matrices, respectively. $\alpha_x = f*m_x$, $\alpha_y = f*m_y$, Here f is the focal length of the camera, m_x, m_y are the scale factors relating pixels to distance, s is skew coefficient between the x and y axis and it is often zero, u_o, v_o are the principal point. According to [34], the projection Matrix P is given by:

$$P = \begin{bmatrix} q_{11} & q_{12} & q_{13} & q_{14} \\ q_{21} & q_{22} & q_{23} & q_{24} \\ q_{31} & q_{32} & q_{33} & q_{34} \end{bmatrix} \tag{5}$$

So the Eq.3 which using to map each pixel on the image plane to a top down view can be re-written as:

$$\begin{bmatrix} u_i \\ v_i \\ 1 \end{bmatrix} = \begin{bmatrix} q_{11} & q_{12} & q_{13} & q_{14} \\ q_{21} & q_{22} & q_{23} & q_{24} \\ q_{31} & q_{32} & q_{33} & q_{34} \end{bmatrix} \begin{bmatrix} X_w \\ Y_w \\ Z_w \\ 1 \end{bmatrix} \tag{6}$$

We now have a position of the object in the transformed image in pixels. In order to convert it into meters, the automatic calibration of the bird's eye view projection system can be used as it is explained in details by [35]. The use of IPM allows the blimp robot to obtain the bird's eye view of the as it is shown on Fig. 3. It allows to remove the perspective effect from the acquired image to weigh each pixel according to its information content. The SURF algorithm might now detect the object with scale invariant or rotation invariant as it is shown in Fig. 4. Then, we could obtain the approximated distances between the camera (blimp robot) and the object. The whole working process of blimp system is shown in Fig. 5.

Fig. 3. IPM transformation for the object

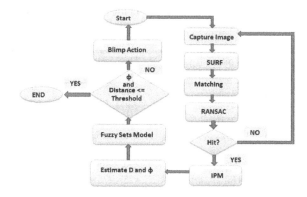

Fig. 4. The SURF feature points

Fig. 5. The blimp system process

4   Possibilities Histograms and Fuzzy Controller

The mostly used controller for the nonlinear system is the fuzzy controller which is able to have robust control of the robot and it has low sensitivity to a variation of parameters or noise levels [36]. In this section the implementation of fuzzy controllers is presented that is based on the visual information previously described; to generate vectorization angle for main propellers as well as yaw angle commands for the blimp. For this work, two fuzzy controllers were implemented and both of them have two inputs and one output. However, the main problem in the fuzzy controller is how to design the fuzzy knowledge base or more precisely the fuzzy membership functions. Therefore, the combination between the possibility theory and fuzzy sets leads to model the complex systems empirically without regard to the presence of the human agent or expert. After we designed the trapezoidal membership functions, we have used bacterial algorithm which allows changing the membership functions breakpoints.

4.1   Yaw Fuzzy Controller

The first input is estimated angle between center of image and center of object. The second input is difference between last two estimations (on last two frames). The output of this controller is the voltage commands for rear motor for changing the heading position of the blimp. Many experiments were done to study the effect of the estimated angles on rear voltage. The angle here is between the center of the image and the center of the projection point of the object on the image ϕ. The general measuring data were collected for different angles, then we analyzed these data to propose fuzzy sets model by using frequency and possibilities distribution. In each case the voltage information were recorded as well as the angle values. A summary of these experiments are given in Table 1. After studying the frequency distributions of these data, we could categorize them into five main groups as it is summarized in Table 2. Then analyzed these groups in order to find the vectors ($\vec{A}$) and random set values (S). The mathematics of the random sets are complicated, but in the finite case which means they take values on subsets of universal Ω they could be seen simply as the following:

$$S = \{< A_i, \delta_i >: \delta_i > 0\} \qquad (7)$$

Where: A is the measuring record subset, δ_i is the evidence function.

Table 1. The empirical data

First Input (Angle)	Rear Voltage	The Vector	The Sets
[-90,-80]	[-1000, -820]	<-1000, -820>	{[-1000, -820]}
[-80,-55]	[-814, -451]	<-814, -451>	{[-814, -451]}
[-55,-30]	[-495, -257]	<-495, -257>	{[-495, -257]}
[-30,-5]	[-281, -15]	<-281, -15>	{[-281, -15]}
[-5,5]	[-10, 6]	<-10, 6>	{[-10, 6]}
[5,30]	[10, 246]	<10, 246>	{[10, 246]}
[30,55]	[257, 478]	<257, 478>	{[257, 478]}
[55,80]	[575, 772]	<575, 772>	{[575, 772]}
[80,90]	[880, 1000]	<880, 1000>	{[880, 1000]}

Table 2. Frequency Distribution of the data

Data	A_i	S_i
[-90,-55]	<[-1000, -820], [-814, -451]>	{[-1000, -820] = 0.5, [-814, -451]= 0.5}
[-55,-5]	<[-495, -257], [-281, -15]>	{[-495, -257] = 0.5, [-281, -15]= 0.5}
[-5, 5]	<[-10, 6]>	{[-10, 6]=1}
[5, 55]	<[10, 246], [257, 478] >	{[10, 246] = 0.5, [257, 478]= 0.5 }
[55, 90]	< [575, 772], [880, 1000]>	{ [575, 772] = 0.5, [880, 1000] = 0.5}

Table 3. Histograms parameters

E^L	E^R	Core	Support
[-1000, -814]	[-820, -451]	[-814, -820]	{(-1000, -814), (-814, -820), (-820, -451)}
[-495, -281]	[-257, -15]	[-257, -281]	{(-495, -281), (-281, -257), (-257, -15)}
[-10, 6]	[-10, 6]	[-10, 6]	{(-10,6)}
[10, 257]	[246, 478]	[246, 257]	{(10, 257), (246,257), (246, 478)}
[575, 880]	[772, 1000]	[772,880]	{(575, 880), (772,880), (772, 1000)}

The form of possibility histogram and distributions (π) depends on the core and support of the measurement record sets as it is shown in Table 3. The core and support of the possibilities distribution for the histogram is given by:

$$C_i(\pi) = \left[E_i^L, E_i^R \right] \tag{8}$$

$$\sup_i(\pi) = \left[E_i^L, E_i^R \right] \tag{9}$$

Where E_L and E_R are the left and right endpoints vectors, respectively. The possibilities histograms for the data as shown in Fig. 6 can be transferred to fuzzy membership functions as shown in Fig. 7 without any changes because that both of them has the same mathematical description and all possibilistic histogram are fuzzy intervals. Fig. 8 also shows the histograms for the first input. The general formula of the above trapezoidal membership functions is given by:

$$\mu_A(x) = \begin{cases} x - a/b - a & a < x < b \\ 1 & b \le x < c \\ d - x/d - c & c < x < d \\ 0 & otherwise \end{cases} \tag{10}$$

The new output of the membership functions can be calculated as the following [37]:

$$\begin{aligned} m_i &= 2d_i - 2a_i + c_i - b_i \\ n_i &= c_i^2 + 2c_id_i + 2d_i^2 - 2a_i^2 - 2a_ib_i - b_i^2 \\ F &= (n_i - m_i^2 - c_im_i)/(2m_i + c_i - b_i) \\ H &= F + m_i \end{aligned} \tag{11}$$

Where a_i, b_i, c_i, and d_i are the breakpoints of the trapezoidal membership functions and F, H are the new breakpoints for left and right sides. The optimized and final membership functions for the output and inputs are shown in Fig. 9, Fig. 10 and Fig. 11 , respectively. Each input and the output has 5 linguistic variables. The number of rules for this controller were 25 base rules.

Fig. 6. The histograms for Output

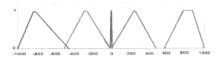

Fig. 7. μ for Output

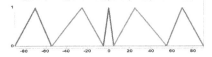

Fig. 8. Histogram for Input 1

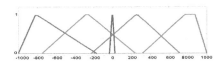

Fig. 9. Optimized Output μ

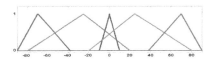

Fig. 10. Optimized μ for Input 1

Fig. 11. μ for Input 2

4.2   Vectorization Fuzzy Controller

For this controller, the first input is estimated distances between blimp and the 3D object target D. The second input is the difference between last two measures. Then, the output is the vectorization angle for the main propellers. Same procedures were done here in order to design the fuzzy membership functions for the vectorization angle. The empirical data ,the analysis as well as the histograms parameters are given in Table 4,5 and 6.

Table 4. The empirical data

First input (Distance)	Vectorization Voltage	The Vector	The Sets
[120,160]	[-232,-98]	<-232,-98>	{[-232,-98]}
[160,200]	[-114,5]	<-114,5>	{[-114,5]}
[200,240]	[-17, 30]	<-17, 30>	{[-17, 30]}
[240,280]	[0, 123]	<0, 123>	{[0, 123]}
[280,320]	[100, 204]	<100, 204>	{[100, 204]}
[320,360]	[228, 317]	<228, 317>	{[228, 317]}
[360,400]	[325, 435]	<325, 435>	{[325, 435]}
[400,440]	[474, 552]	<474, 552>	{[474, 552]}
[440,480]	[664, 720]	<664, 720>	{[664, 720]}

Table 5. Frequency Distribution of the data

Data	A_i	S_i
[120,200]	<[-232,-98], [-114,5]>	{[-232,-98] = 0.5, [-114,5]=0.5}
[200,240]	<[-17, 30]>	{[-17, 30] = 1}
[240,320]	<[0, 123], [100, 204] >	{[0, 123] = 0.5, [100, 204]= 0.5}
[320, 400]	<[228, 317], [325, 435]>	{[228, 317]= 0.5, [325, 435] = 0.5}
[400, 480]	<[474, 552], [664, 720]>	{[474, 552] = 0.5, [664, 720] = 0.5}

Table 6. Histograms parameters

E_L	E_R	Core	Support
[-232, -114]	[-98, 5]	[-98,-114]	{(-232, -114), (-98,-114), (-98, 5)}
[-17, 30]	[-17, 30]	[-17, 30]	{(-17, 30), (-17, 30), (-17, 30)}
[0, 100]	[123, 204]	[123, 100]	{(0, 100), (123, 100), (123, 204)}
[228, 325]	[317, 435]	[317,325]	{(228, 325), (317,325), (317, 435)}
[474, 664]	[552, 720]	[552, 664]	{(474, 664), (552, 664), (552, 720)}

Fig. 12. Histograms for Output

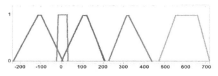

Fig. 13. μ for Output

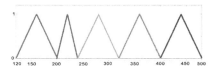

Fig. 14. The Histograms for Input1

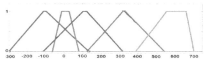

Fig. 15. The Final Output μ

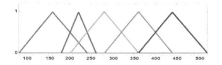

Fig. 16. The Final μ for Input1

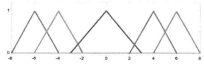

Fig. 17. μ for Input 2

Same procedure for this controller was done in order to find the histograms for output and input1 as they are shown in Fig. 12, Fig. 13 and Fig. 14. Then, by using bacterial algorithm the membership functions have been obtained as shown in Fig. 15, Fig. 16 and Fig. 17. Therefore, 5 linguistic variables for each input as well as 5 for input had been found. The total number of the fuzzy rules were 21 rules.

5   Experimental Results

In order to verify the proposed fuzzy vision system, several experiments of the complete system were conducted. During these experiments the blimp robot was flying at a certain altitude as well as the 3D objects move in the plane with constant altitude. We assume that the 3D object target is already in the view of the on-board camera. The target would be identified and tracked in the video sequence by the vision system. Based on the vision data, the vectorization angle for the main propellers and the yaw angle were controlled to follow the target and keep it in a certain position in the image as well as in a certain distance from the blimp. Fig. 18 shows the distance between the blimp and the object. We assume here that the target is static at a certain position from the blimp. The blimp vision system detects and flies toward the target. Then, the engine will stop and the blimp stop moving at a certain distance which is here 75 cm. In Fig. 19, Fig. 20 and Fig. 21, they show the blimp behavior in order to control the yaw angle and keep the object at the center of the image plane. The distance in pixels between the blimp and the target is shown in Fig. 22. The vectorization angle between the blimp robot and the object is shown in Fig. 23 and we should note that the blimp controller try to follow the object as well as keep the blimp at a certain altitude which means the vectorization angle should back to 90 Degree during the mission, otherwise the blimp will go down while it is flying. In Fig. 24 shows the trajectory which has been made by the blimp robot during the mission in our LAB. These data are collected using the IMU mounted on the blimp robot , then, reconstructed after the missions are completed. Also, the sequences images for the experiments is shown in Fig. 25.

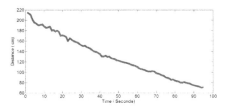
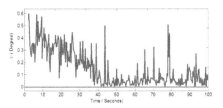

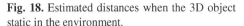

Fig. 18. Estimated distances when the 3D object static in the environment.

Fig. 19. The angle between the center and the object in image plane

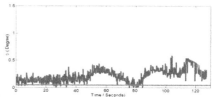

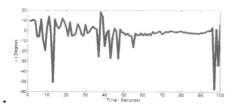

Fig. 20. The angle between the center and the object in image plane

Fig. 21. The angle between the center and the object in image plane

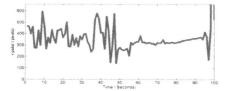

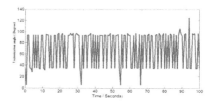

Fig. 22. distance in polar coordinate (in pixels)

Fig. 23. Vectorization angle

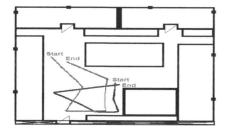

Fig. 24. Trajectory of blimp during tests

Fig. 25. Experiments Sequences images

6   Conclusion

In this paper, an efficient vision-based object detection and localization algorithm to realize the autonomous 3D object following has been presented. The visual information is provided by SURF algorithm as well as Inverse Perspective Mapping. The fuzzy logic controllers have been designed experimentally in order to keep the blimp at a certain distance and maintain it in the center of the image. The experiments results validate that the algorithm is not only be able to track the target effectively, but also improves the robustness and accuracy. Also, we presented a method in order to estimate the distance between two 3D objects in the space based on visual information and using single camera. This estimation helps to localize the 3D object with respect to the known blimp position, assuming the altitude of both are known. However, this algorithm could face some limitations due to the vision sensor characteristics as well as the size of the target that could affect the performance. In the future, the multi 3D target tracking using binary sensor network will be presented to estimate the locations of two flying objects on indoor environments, takes into account that the targets are flying at unknown altitude. Also, the Extended Kalman filter (EKF) or modified EKF might be used for extracting necessary information about the dynamics of the moving objects.

References

[1] Browning, B., Veloso, M.: Real Time, Adaptive Color Based Robot Vision. In: IEEE/RSJ International Conference on Intelligent Robots and Systems, pp. 3871–3876 (2005)

[2] Guenard, N., Hamel, T., Mahony, R.: A Practical Visual Servo Control for an Unmanned Aerial Vehicle. IEEE Trans. Robot. 24, 331–340 (2008)

[3] Lin, F., Lum, K., Chen, B., Lee, T.: Development of a Vision-Based Ground Target Detection and Tracking System for a Small Unmanned Helicopter. Sci. China—Series F: Inf. Sci. 52, 2201–2215 (2009)

[4] Korodi, A., Codrean, A., Banita, L., Volosencu, C.: Aspects Regarding The Object Following Control Procedure for Wheeled Mobile Robots. WSEAS Trans. System Control 3, 537–546 (2008)

[5] Betser, A., Vela, P., Pryor, G., Tannenbaum, A.: Flying Information Using a Pursuit Guidance Algorithm. In: American Control Conference, vol. 7, pp. 5085–5090 (2005)

[6] Savkin, A., Teimoori, H.: Bearings-Only Guidance of an Autonomous Vehicle Following a Moving Target With a Smaller Minimum Turning Radius. In: 47th IEEE Conference on Decision and Control, pp. 4239–4243 (2008)

[7] Fukao, T., Kanzawa, T., Osuka, K.: Tracking Control of an Aerial Blimp Robot Based on Image Information. In: 16th IEEE International Conference on Control Applications part of IEEE Multi-Conference on Systems and Control, Singapore, pp. 874–879 (2007)

[8] Canals, R., Roussel, A., Famechon, J., Treuillet, S.: A Bi-processor Oriented Vision-Based Target Tracking System. IEEE Trans. Ind. Electron. 49(2), 500–506 (2002)

[9] Mejias, L., Saripalli, S., Cervera, P., Sukhatme, G.: Visual Servoing of an Autonomous Helicopter in Urban Areas Using Feature Tracking. Journal of Field Robot 23, 185–199 (2006)

[10] Hu, W.M., Tan, T.N., Wang, L., Maybank, S.: A Survey on Visual Surveillance of Object Motion and Behaviours. IEEE Trans. Syst. 34, 334–352 (2004)

[11] Hu, Y., Zhao, W., Wang, L.: Vision-based Target Tracking and Collision Avoidance for Two Autonomous Robotic Fish. IEEE Trans. Ind. Electron. 56, 1401–1410 (2009)

[12] Xu, D., Han, L., Tan, M., Li, Y.: Ceiling-Based Visual Positioning for an Indoor Mobile Robot With Monocular Vision. IEEE Trans. Ind. Electron. 56, 1617–1628 (2009)

[13] Lowe, D.: Distinctive Image Features From Scale Invariant Keypoints. International Journal of Computer Vision 60(2), 91–110 (2004)

[14] Bay, H., Ess, A., Tuyelaars, T., Gool, V.: SURF: Speeded Up Robust Features. Computer Vision and Image Understanding CVIU 110(3), 346–359 (2008)

[15] Saripalli, S., Sukhatme, G.S.: Landing a helicopter on a moving target. In: Proceedings of IEEE International Conference on Robotics and Automation, Rome, Italy, pp. 2030–2035 (2007)

[16] Bourquardez, O., Mahony, R., Guenard, N., Chaumette, F., Hamel, T., Eck, L.: Image based visual servo control of the translation kinematics of a quadrotor aerial vehicle. IEEE Transactions on Robotics 25(3), 743–749 (2009)

[17] Choi, H., Kim, Y.: UAV guidance using a monocular-vision sensor for aerial target tracking. Control Engineering Practice Journal 22, 10–19 (2014)

[18] Passerieus, J.M., Cappel, D.V.: Optimal Observer Maneuver for Bearings-Only Tracking. IEEE Transactions on Aerospace and Electronic Systems 34(3), 777–788 (1998)

[19] Watanabe, Y., Johnson, E.N., Calise, A.J.: Optimal 3D Guidance from a 2D Vision Sensor. In: AIAA Guidance, Navigation, and Control Conference, Providence, RI (2004)

[20] Cao, C., Hovakimyan, N.: Vision-Based Aerial Tracking using Intelligent Excitation. In: American Control Conference, Portland, OR, pp. 5091–5096 (2005)

[21] Stepanyan, V., Hovakimyan, N.: A Guidance Law for Visual Tracking of a Maneuvering Target. In: IEEE Proceeding of American Control Conference, Minneapolis, MN, pp. 2850–2855 (2006)

[22] Stepanyan, V., Hovakimyan, N.: Adaptive Disturbance Rejection Controller for Visual Tracking of a Maneuvering Target. Journal of Guidance, Control and Dynamics 30(4), 1090–1106 (2007)

[23] Mondragon, I.F., Campoy, P., Olivares-Mendez, M.A., Martinez, C.: 3D Object following based on visual information for unmanned aerial vehicles. In: Robotics Symposium IEEE IX Latin American and IEEE Colombian Conference on Automatic Control and Industry Applications, Bogota (2011)

[24] Olivares-Mendez, M.A., Mondragon, I.F., Campoy, P., Mejias, L., Martinez, C.: Aerial Object Following Using Fuzzy Servoing. In: Proceeding of Workshop on Research, Development and Education on Unmanned Aerial Systems RED-UAS, Seville, Spain (2011)

[25] Joslyn, C.: In Support of an Independent Possibility Theory. In: de Cooman, G., Raun, D., Kerre, E.E. (eds.) Foundations and Applications of Possibility Theory, pp. 152–164. World Scientific, Singapore (1995a)

[26] Zadeh, L.: Fuzzy Sets as the Basis for a Theory of Possibility. Fuzzy Sets and Systems 1, 3–28 (1978); Reprinted in Fuzzy Sets and Systems 100(suppl.), 9–34 (1999)

[27] Dubois, D., Prade, H.: Probability Theory and Multiple-valued Logics. A Clarification. Annals of Mathematics and Artificial Intelligence 32, 35–66 (2001)

[28] Joslyn, C.: Measurement of Possibilistic Histograms from Interval Data. NCR Research Associate, Mail Code 522, NASA Goddard Space Flight Center, Greenbelt, MD 20771, USA (1996)

[29] Al-Jarrah, R., Roth, H.: Design Blimp Robot based on Embedded System & Software Architecture with high level communication & Fuzzy Logic. In: Proceeding in IEEE 9th International Symposium on Mechatronics & its Applications (ISMA 2013), Amman, Jordan (2013)

[30] Al-Jarrah, R., Roth, H.: Developed Blimp Robot Based on Ultrasonic Sensors Using Possibilities Distributions and Fuzzy Logic. In: 5th International Conference on Computer and Automation Engineering, ICCAE, Belgium, vol. 1(2), pp. 119–125 (2013)

[31] Al-Jarrah, R., AitJellal, R., Roth, H.: Blimp based on Embedded Computer Vision and Fuzzy Control for Following Ground Vehicles. In: 3rd IFAC Symposium on Telematics Applications (TA 2013), Seoul, Korea (2013)

[32] Fischer, M.A., Bolles, R.C.: Random Sample Consensus: a paradigm for Model Fitting With Applications to Image Analysis and Automated Cartography. Communication of the ACM 24(6), 381–395 (1981)

[33] Mallot, H.A., Biilthoff, H.H., Little, J.J., Bohrer, S.: Inverse Perspective Mapping Simplifies Optical Flow Computation and Obstacle Detection. Biological Cybernetics, 177–185 (1991)

[34] Hartley, R., Zisserman, A.: Multiple View Geometry in Computer Vision. Cambridge University Press (2000)

[35] Bradsky, G., Kaebler, A.: Learning OpenCV: Computer Vision with the OpenCV Library, pp. 408–412. Oreilly Media (2008)

[36] Jang, J.-S.R., Sun, C.-T., Mizutani, E.: Nuero Fuzzy and Soft Computing, pp. 13–91. Prentice-Hall, Upper Saddle River (1997)

[37] Botzheim, J., Hámori, B., Kóczy, L.T.: Applying Bacterial Algorithm to Optimize Trapezoidal Membership Functions in a Fuzzy Rule Base. In: Proceeding of the International Conference on Computational Intelligence, Theory and Applications, 7th Fuzzy Days, Dortmund, Germany (2001)

At Odds with Curious Cats, Curious Robots Acquire Human-Like Intelligence

Dominik M. Ramík, Kurosh Madani, and Christophe Sabourin

Signals, Images, and Intelligent Systems Laboratory (LISSI / EA 3956), University Paris-Est Creteil, Senart-FB Institute of Technology, 36-37 rue Charpak, 77127 Lieusaint, France
{dominik.ramik,madani,sabourin}@u-pec.fr

Abstract. This work contributes to the development of a real-time intelligent system allowing to discover and to learn autonomously new knowledge about the surrounding world by semantic interaction with human. Based on human's curiosity mechanism, the learning is accomplished by observation and by interaction with human. We provide experimental results implementing the approach on a humanoid robot in a real-world environment including every-day objects. We show, that our approach allows a humanoid robot to learn without negative input and from small number of samples.

Keywords: intelligent system, visual saliency, autonomous learning, learning by interaction.

1   Introduction

If nowadays machines and robotic bodies are fully automated outperforming human capacities, nonetheless, none of them can be called truly intelligent or can defeating human's cognitive skills. The fact that human-like machine-cognition is still beyond the reach of contemporary science only proves how difficult the problem is. Somewhat, it is due to the fact that we are still far from fully understanding the human cognitive system. Partly, it is so because if contemporary machines are often fully automatic, they linger rarely fully autonomous in their knowledge acquisition. Nevertheless, the concepts of bio-inspired or human-like machine-cognition remain foremost sources of inspiration for achieving intelligent systems (intelligent machines, intelligent robots, etc...).

Emergence of cognitive phenomena in machines has been and remains active part of research efforts since the rise of Artificial Intelligence (AI) in the middle of the last century. Among others, [1] provides a survey on cognitive systems. It accounts on different paradigms of cognition in artificial agents markedly on the contrast of emergent versus cognitivist paradigms and on their hybrid combinations. The work of [2] brings an in-depth review on a number of existing cognitive architectures such those which adheres to the symbolic theory and reposes on the assumption that human knowledge can be divided to two kinds: declarative and procedural. Another discussed architecture belongs to class of those using "If-Then" deductive rules dividing knowledge again on two kinds: concepts and skills. In contrast to above-mentioned

V. Golovko and A. Imada (Eds.): ICNNAI 2014, CCIS 440, pp. 112–123, 2014.

works, the work of [3] focuses the area of research on cognition and cognitive robots discussing purposes linking knowledge representation, sensing and reasoning in cognitive robots. However, there is no cognition without perception (a cognitive system without the capacity to perceive would miss the link to the real world and so it would be impaired) and thus autonomous acquisition of knowledge from perception is a problem that should not be skipped when dealing with cognitive systems.

Prominently to the machine-cognition's issue is the question: "what is the compel or the motivation for a cognitive system to acquire new knowledge?" For human cognitive system Berlyne states, that it is the curiosity that is the motor of seeking for new knowledge [4]. Consequently a few works have been since there dedicated to incorporation of curiosity into a number of artificial systems including embodied agents or robots. However the number of works using some kind of curiosity motivated knowledge acquisition with implementation to real agents (robots) is still relatively small. Often authors view curiosity only as an auxiliary mechanism in robot's exploration behavior. One of early implementations of artificial curiosity may be found in [5]. Accordingly to the author, the introduction of curiosity further helps the system to actively seek similar situations in order to learn more. On the field of cognitive robotics a similar approach may be found in [6] where authors present an approach including a mechanism called "Intelligent Adaptive Curiosity". Two experiments with AIBO robot are presented showing that the curiosity mechanism successfully stimulates the learning progress. In a recent publication, authors of [7] implement the psychological notion of surprise-curiosity into the decision making process of an agent exploring an unknown environment. Authors conclude that the surprise-curiosity driven strategy outperformed classical exploration strategy regarding the time-energy consumed in exploring the delved environment. On the other hand, the concept of surprise, relating closely the notion of curiosity, has been exploited in [8] by a robot using the surprise in order to discover new objects and acquire their visual representations. Finally, the concept of curiosity has been successfully used in [9] for learning affordances of a mobile robot in navigation task. The mentioned works are attempting to respond the question: "how an autonomous cognitive system should be designed in order to exhibit the behavior and functionality close to its human users".

That is why even though in English literature "curiosity killed a cat" (BENNY— (with a wink): "Curiosity killed a cat! Ask me no questions and I'll tell you no lies.", Different, Eugene O'Neill, 1920), taking into consideration the aforementioned enticing benefits of curiosity, we have made it our principle foundation in investigated concept. The present paper is devoted to the description of a cognitive system based on artificial curiosity for high-level human-like knowledge acquisition from visual information. The goal of the investigated system is to allow the machine (such as a humanoid robot) to observe, to learn and to interpret the world in which it evolves, using appropriate terms from human language, while not making use of a priori knowledge. This is done by word-meaning anchoring based on learning by observation stimulated (steered) by artificial curiosity and by interaction with the human. Our model is closely inspired by juvenile learning behavior of human infants ([10] and [11]).

In Section 2, we detail our approach by outlining its architecture and principles, we explain how beliefs about the world are generated and evaluated by the system and we describe the role of human-robot interaction in the learning process. Section 3 focuses the validation of the proposed approach on a real robot in real world. Finally Section 4 discusses the achieved results and outlines the future work.

2   Artificial Curiosity and Learning-By-Interaction

Accordingly to Berlyne's theory of human curiosity [4], two cognitive levels contribute to human's desire of acquiring new knowledge. The first is so-called "perceptual curiosity", which leads to increased perception of stimuli. It is a lower level cognitive function, more related to perception of new, surprising or unusual sensory input. It contrasts to repetitive or monotonous perceptual experience. The other one is called "epistemic curiosity", which is more related to the "desire for knowledge that motivates individuals to learn new ideas, eliminate information-gaps, and solve intellectual problems" [12]. It also seems that it acts to stimulate long-term memory in remembering new or surprising (e.g. what may be contrasting with already learned) information [13]. By observing the state of the art (including the referenced ones), it may be concluded that the curiosity is usually used as an auxiliary mechanism instead of being the fundamental basis of the knowledge acquisition. To our best knowledge there is no work to date which considers curiosity in context of machine cognition as a drive for knowledge acquisition on both low (perceptual) level and high ("semantic") level of the system. Without striving for biological plausibility whilst by analogy with natural curiosity, we founded our system on two cognitive levels ([14], [15]). The first ahead of reflexive visual attention plays the role of perceptual curiosity and the second coping with intentional learning-by-interaction undertakes the role of epistemic curiosity.

2.1   From Observation to Interpretation

The problem of autonomous learning conveys the inbuilt problem of distinguishing the pertinent sensory information from the impertinent one. The solution to this task is natural for human, it remain very far from being obvious for a robot. In fact, when a human points to one object among many others giving a description of that pointed object using his human natural language, the robot still has to distinguish, which of the detected features and perceived characteristics of the object the human is referring to. To achieve correct anchoring, the proposed architecture adopts the following strategy. By using its perceptual curiosity, realized thanks to artificial salient vision and adaptive visual attention (described in [16], [17] and [18]), the robot extracts features from important objects found in the scene along with the words the human used to describe the objects. Then, the robot generates its beliefs about which words could describe which features. Using the generated beliefs as organisms in a genetic algorithm, the robot determines its "most coherent belief". To calculate the fitness, a classifier is trained and used to interpret the objects the robot has already seen. The utterances pronounced by the human for each object are compared with those the robot would use to describe it based on its current belief. The closer the robot's description is to that given by the human, the higher the fitness is. Once the evolution has been finished, the belief with the highest fitness is adopted by the robot and is used to interpret occurrences of new (unseen) objects. Fig. 1 depicts through an example important parts and operations of the proposed system.

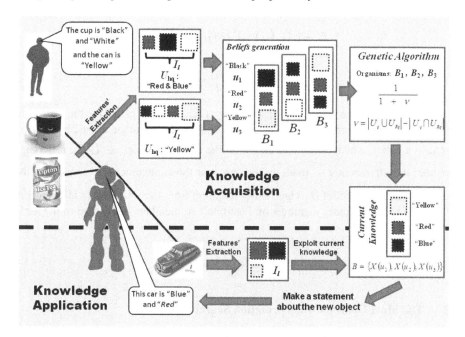

Fig. 1. Example showing main parts of the system's operation in the case of autonomous learning of colors

Let us suppose a robot equipped by a sensor observing the surrounding world and interacting with the human. The world is represented as a set of features $I = \{i_1, i_2, \cdots, i_k\}$, which can be acquired by robot's sensor. Each time the robot makes an observation o, its epistemic curiosity stimulates it to interact with the human asking him to gives a set of utterances U_H describing the found salient objects. Let us denote the set of all utterances ever given about the world as U. The observation o is defined as an ordered pair $o = \{I_i, U_H\}$, where $I_i \subseteq I$, expressed by (1), stands for the set of features obtained from observation and $U_H \subseteq U$ is the set of utterances (describing O) given by human in the context of that observation. i_p denotes the pertinent information for a given u (i.e. features that can be described semantically as u in the language used for communication between the human and the robot), i_i the impertinent information (i.e. features that are not described by the given u, but might be described by another $u_i \in U$) and sensor noise ε. The goal is to distinguish the pertinent information from the impertinent one and to correctly map the utterances to appropriate perceived stimuli (features). Let us define an interpretation $X(u) = \{u, I_j\}$ of an utterance u as an ordered pair where $I_j \subseteq I$ is a set of features from I. So, the belief B is defined accordingly to (2) as an ordered set of $X(u)$ interpreting utterances u from U.

$$I_l = \bigcup_{U_H} i_p(u) + \bigcup_{U_H} i_l(u) + \varepsilon \qquad (1)$$

$$B = \{X(u_1), \cdots, X(u_n)\} \text{ with } n = |U| \qquad (2)$$

Accordingly to the criterion expressed by (3), one can calculate the belief B which interprets coherently the observations made so far: in other words, by looking for such a belief, which minimizes across all the observations $o_q \in O$ the difference between the utterances U_{Hq} made by human, and those utterances U_{Bq}, made by the system by using the belief B. Thus, B is a mapping from the set U to I: all members of U map to one or more members of I and no two members of U map to the same member of I.

$$\arg\min_B \left(\sum_{q=1}^{|O|} |U_{Hq} - U_{Bq}| \right) \qquad (3)$$

2.2   The Most Coherent Interpretation Search

Although the interpretation's coherence is worked out by computing the belief B accordingly to equation (3), the system has to look for a belief B, which would make the robot describing a particular scene with utterances as close and as coherent as possible to those that a human would made on the same (or similar) scene. For this purpose, instead performing the exhaustive search over all possible beliefs, we propose to search for a suboptimal belief by means of a Genetic Algorithm (GA). For doing that, we assume that each organism within it has its genome constituted by a belief, which, results into genomes of equal size $|U|$ containing interpretations $X(u)$ of all utterances from U.

In our genetic algorithm, the genomes' generation is a belief generation process generating genomes (e.g. beliefs) as follows. For each interpretation $X(u)$ the process explores whole the set O. For each observation $o_q \in O$, if $u \in U_{Hq}$ then features $i_q \in I_q$ (with $I_q \subseteq I$) are extracted. As described in (1), the extracted set of features contains as well pertinent as impertinent features. The coherent belief generation is done by deciding, which features $i_q \in I_q$ may possibly be the pertinent ones. The decision is driven by two principles. The first one is the principle of "proximity", stating that any feature i is more likely to be selected as pertinent in the context of

u, if its distance to other already selected features is comparatively small. The second principle is the "coherence" with all the observations in O. This means, that any observation $o_q \in O$, corresponding to $u \in U_{Hq}$, has to have at least one feature assigned into I_q of the current $X(u) = \{u, I_q\}$.

To evaluate a given organism, a classifier is trained, whose classes are the utterances from U and the training data for each class $u \in U$ are those corresponding to $X(u) = \{u, I_q\}$, i.e. the features associated with the given u in the genome. This classifier is used through whole set O of observations, classifying utterances $u \in U$ describing each $o_q \in O$ accordingly to its extracted features. Such a classification results in the set of utterances U_{Bq} (meaning that a belief B is tested regarding the qth observation). The fitness function evaluating the fitness of each above-mentioned organism is defined as "disparity" between U_{Bq} and U_{Hq} (defined in previous subsection) which is computed accordingly to the equation (4), where v is the number of utterances that are not present in both sets U_{Bq} and U_{Hq} (e.g. either missed or are superfluous utterances interpreting the given features). The globally best fitting organism is chosen as the belief that best explains observations O made (by robot).

$$D(v) = \frac{1}{1+v} \quad \text{with} \quad v = \left| U_{Hq} \bigcup U_{Bq} \right| - \left| U_{Hq} \bigcap U_{Bq} \right| \tag{4}$$

It is important to note that here the above-described GA based evolutionary process doesn't operates as only an optimizer but it generate the machines (e.g. robot's) most coherent belief about the observation accomplished by this robot and about the way that the same robot will autonomously construct a human-like description of the observed reality. In other words, it is the GA based evolutionary process that drives the robot's most coherent semantic understanding of the observed reality. It plays also a key role in implementation of the epistemic curiosity because the drop of the search for the most coherent belief, due to leakage of knowledge about the observed reality, makes the robot interacting with its human counterpart and thus drives its epistemic curiosity.

2.3   Role of Human-Robot Interaction

Human beings learn both by observation and by interaction with the world and with other human beings. The former is captured in our system in the "best interpretation search" outlined previous subsections. The latter type of learning requires that the robot be able to communicate with its environment and is facilitated by learning by observation, which may serve as its bootstrap. In our approach, the learning by interaction is carried out in two kinds of interactions: human-to-robot and robot-to-human. The human-to-robot interaction is activated anytime the robot interprets wrongly the world. When the human receives a wrong response (from robot), he provides the robot a new observation by uttering the desired interpretation. The robot takes this new corrective knowledge about the world into account and searches for a new interpretation of the world conformably to this new observation. The robot-to-human interaction may be activated when the robot attempts to interpret a particular feature classified with a very low confidence: a sign that this feature is a borderline example. In this case, it may be beneficial to clarify its true nature. Thus, led by the epistemic curiosity, the robot asks its human counterpart to make an utterance about the uncertain observation. If the robot's interpretation is not conforming to the utterance given by the human (robot's interpretation was wrong), this observation is recorded as a new knowledge and a search for the new interpretation is started.

3   Implementation and Validation on Real Robot

The designed system has been implemented on NAO robot (from Aldebaran Robotics). It is a small humanoid robot which provides a number of facilities such as onboard camera (vision), communication devices and onboard speech generator. The fact that the above-mentioned facilities are already available offers a huge save of time, even if those faculties remain quite basic in that kind of robots. If NAO robot integrates an onboard speech-recognition algorithm (e.g. some kind of speech-to-text converter) which is sufficient for "hearing" the tutor, however its onboard speech generator is a basic text-to-speech converter. It is not sufficient to allow the tutor and the robot conversing in natural speech. To overcome NAO's limitations relating this purpose, the TreeTagger tool [1] was used in combination with robot's speech-recognition system to obtain the part-of-speech information from situated dialogs. Standard English grammar rules were used to determine whether the sentence is demonstrative (e.g. for example: "This is an apple."), descriptive (e.g. for example: "The apple is red.") or an order (e.g. for example: "Describe this thing!"). To communicate with the tutor, the robot used its text-to-speech engine.

3.1   Implementation

The core of the implementation's architecture is split into five main units: Communication Unit (CU), Navigation Unit (NU), Low-level Knowledge Acquisition Unit (LKAU), High-level Knowledge Acquisition Unit (HLAU) and Behavior Control Unit (BCU). Fig. 2 illustrates the bloc-diagram of the implementation's architecture. The aforementioned units control NAO robot (symbolized by its sensors, its actuators and its interfaces in Fig. 2) through its already available hardware and software facilities. In other words, the above-mentioned architecture controls the whole robot's behavior.

The purpose of NU is to allow the robot to position itself in space with respect to objects around it and to use this knowledge to navigate within the surrounding environment. Capacities needed in this context are obstacle avoidance and

---

[1] Developed by the ICL at University of Stuttgart, available online at: http://www.ims
.uni-stuttgart.de/projekte/corplex/TreeTagger

determination of distance to objects. Its sub-unit handling spatial orientation receives its inputs from the camera and from the LKAU. To get to the bottom of the obstacle avoidance problem, we have adopted a technique based on ground color modeling. Inspired by the work presented in [19], color model of the ground helps the robot to distinguish free-space from obstacles.

The LKAU ensures gathering of visual knowledge, such as detection of salient objects and their learning (by the sub-unit in charge of salient object detection) and sub-recognition (see [18] and [20]). Those activities are carried out mostly in an "unconscious" manner, i.e. they are run as an automatism in "background" while collecting salient objects and learning them. The learned knowledge is stored in Long-term Memory for further use.

The HKAU is the center where the intellectual behavior of the robot is constructed. Receiving its features from the LKAU (visual features) and from the CU (linguistic features), this unit processes the beliefs' generation, the most coherent belief's emergence and constructs the high-level semantic representation of acquired visual knowledge. Unlike the LKAU, this unit represents conscious and intentional cognitive activity. In some way, it operates as a baby who learns from observation and from verbal interaction with adults about what he observes developing in this way his own representation and his own opinion about the observed world [21].

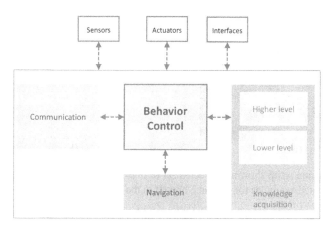

Fig. 2. Bloc diagram of the implementation's architecture

The CU is in charge of robots communication. It includes an output communication channel and an input communication channel. The output channel is composed of a Text-To-Speech engine which generates human voice through loud-speakers. It receives the text from the BCU. The input channel takes its input from a microphone and through an Automated Speech Recognition engine (available in NAO) the syntax and semantic analysis (designed and incorporated in BCU) it provides the BCU labeled chain of strings representing the heard speech.

The BCU plays the role of a coordinator of robot's behavior. It handles data flows and issues command signals for other units, controlling the behavior of the robot and its suitable reactions to external events (including its interaction with humans). BCU received its inputs from all other units and returns its outputs to each concerned unit including robot's devices (e.g. sensors, actuators and interfaces) [21]. The human-robot interaction is performed by this unit in cooperation with HLAU. In other words, driven by HLAU, a part of the robot's epistemic curiosity based behavior is handled by BCU.

3.2   Experimental Validation

The total of 25 every-day objects was collected for purposes of the experiment (Fig.3-a). The collected set has been randomly divided into two sets for training and for testing. The learning set objects were placed around the robot and then a human tutor pointed to each of them calling it by its name. Using its 640x480 monocular color camera, the robot discovered and learned the objects around it by the salient object detection approach we have described in [18]. Here, this approach has been extended by detecting the movement of the human's hand to achieve joint attention. In this way, the robot was able to determine what object the tutor is referring to and to learn its name. Fig. 3-b shows that only a few of exemplars are enough for a correct learning of an item. Fig. 4 shows an example of the learning sequence.

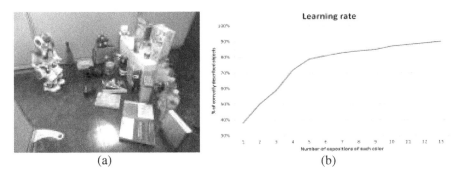

(a)   (b)

Fig. 3. Experimental setup showing collected set of objects (a) and the learning rate versus the number of presented examples (of the same object) to the robot (b)

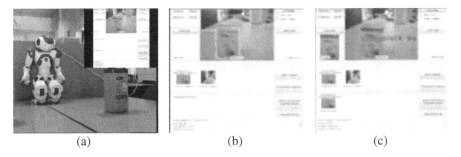

(a)   (b)   (c)

Fig. 4. Example of a learning sequence where robot learns a "yellow chocolate box" (a), detection of that object as a salient object (b) and labeling of that object after learning (c)

Fig. 5. Images from a video sequence showing the robot exploring the room (left-side picture) and finding salient objects in that room (middle and right-side pictures, respectively)

Fig. 6. Images from a video sequence showing the robot interacting with the human and learning-by-interaction from him

Fig. 7. Images from a video sequence showing the robot finding the required book (left-side picture) and how the robot perceives that required book (right-side picture)

Within the above-mentioned experimental protocol, the robot has been asked to explore an unknown environment (an unseen room) where a subset among the 25 collected objects has been placed. After exploration, the human show, one-by-one, objects detected by robot in that room accordingly to its perceptual curiosity and asked the robot if the robot has seen each indicated object. However, the human show also some unseen objects (as a blue teddy-bear). The epistemic curiosity of the robot stimulates it to ask the human additional information and learn-by-interaction with the human the discovered objects. Extracted from the video, Fig. 5 and Fig. 6 show photographs of the robot during the above-described experiment. Here-bellow an example of the Human-Robot interactive learning is reported:

- **Human** [*giving the order to the Robot*]: "Go explore the room."
- **Robot**: "I am exploring the room."

- **Robot** [*after finishing the exploration of the room*]: "I finished the exploring and I have seen several interesting objects."
- **Human** [showing the book to the Robot]: "Have you seen this?"
- **Robot**: "Yes! I have seen this, what is its name?"
- **Human**: "This is a book."
- **Robot**: "OK! I will remember that it is a book."
- **Human** [*taking out from his pocket an unseen blue teddy-bear*]: "Have you seen this?"
- **Robot**: "No! I haven't seen this before, what is this?"
- **Human**: "This is a teddy-bear."
- **Robot**: "OK! I will remember that this is a teddy-bear."

After having learned objects (among which a black book), the robot has been asked to search for the "book" placed in different positions in that room. Fig. 7 shows photographs of the robot during the above-described experiment. Robot searches and successfully finds the book in different positions and stances. Additional experimental results are available on: http://youtu.be/W5FD6zXihOo.

4   Conclusion and Further Work

In this paper, we have presented and validated an intelligent system for high-level knowledge acquisition from observation. Founded on human's curiosity mechanism, the presented system allow to learn in an autonomous manner new knowledge about the surrounding world and to complete (enrich or correct) it by interacting with a human in natural and coherent way. Experimental results using the NAO robot show the pertinence of the investigated concepts.

Several appealing perspectives are pursuing to push further the presented work. The current implemented version allows the robot to work with a single category or property at a time (e.g. for example the color in utterances like "it is red"). We are working on extending its ability to allow the learning of multiple categories at the same time and to distinguish which of the used words are related to which category. While, concerning the middle-term perspectives of this work, they will focus aspects reinforcing the autonomy of such cognitive robots. The ambition here is integration of the designed system to a system of larger capabilities realizing multi-sensor artificial machine-intelligence. There, it will play the role of an underlying part for machine cognition and knowledge acquisition.

References

1. Vernon, D., Metta, G., Sandini, G.: A Survey of Artificial Cognitive Systems: Implications for the Autonomous Development of Mental Capabilities in Computational Agents. IEEE Transactions on Evolutionary Computation 11(2), 151–180 (2007)
2. Langley, P., Laird, J.E., Rogers, S.: Cognitive architectures: Research issues and challenges. Cognitive Systems Research 10(2), 141–160 (2009)
3. Levesque, H.J., Lakemeyer, G.: Cognitive robotics. In: Handbook of Knowledge Representation. Schloss Dagstuhl - Leibniz-Zentrum fuer Informatik, Dagstuhl (2010)

4. Berlyne, D.E.: A theory of human curiosity. British Journal of Psychology 45(3), 180–191 (1954)
5. Schmidhuber, J.: Curious model-building control systems. In: Proceedings of International Joint Conference on Neural Networks (IEEE-IJCNN 1991), vol. 2, pp. 1458–1463 (1991)
6. Oudeyer, P.-Y., Kaplan, F., Hafner, V.V.: Intrinsic Motivation Systems for Autonomous Mental Development. IEEE Transactions on Evolutionary Computation 11(2), 265–286 (2007)
7. Macedo, L., Cardoso, A.: The exploration of unknown environments populated with entities by a surprise-curiosity-based agent. Cognitive Systems Research 19-20, 62–87 (2012)
8. Maier, W., Steinbach, E.G.: Surprise-driven acquisition of visual object representations for cognitive mobile robots. In: Proc. of IEEE Int. Conf. on Robotics and Automation, Shanghai, pp. 1621–1626 (2011)
9. Ugur, E., Dogar, M.R., Cakmak, M., Sahin, E.: Curiosity-driven learning of traversability affordance on a mobile robot. In: Proc. of IEEE 6th Int. Conf. on Development and Learning, pp. 13–18 (2007)
10. Yu, C.: The emergence of links between lexical acquisition and object categorization: a computational study. Connection Science 17(3-4), 381–397 (2005)
11. Waxman, S.R., Gelman, S.A.: Early word-learning entails reference, not merely associations. Trends in Cognitive Science (2009)
12. Litman, J.A.: Interest and deprivation factors of epistemic curiosity. Personality and Individual Differences 44(7), 1585–1595 (2008)
13. Kang, M.J.J., Hsu, M., Krajbich, I.M., Loewenstein, G., McClure, S.M., Wang, J.T.T., Camerer, C.F.: The wick in the candle of learning: epistemic curiosity activates reward circuitry and enhances memory. Psychological Sci. 20(8), 963–973 (2009)
14. Madani, K., Sabourin, C.: Multi-level cognitive machine-learning based concept for human-like artificial walking: Application to autonomous stroll of humanoid robots. Neurocomputing, 1213–1228 (2011)
15. Ramik, D.-M., Sabourin, C., Madani, K.: From Visual Patterns to Semantic Description: a Cognitive Approach Using Artificial Curiosity as the Foundation. Pattern Recognition Letters 34(14), 1577–1588 (2013)
16. Ramik, D.-M., Sabourin, C., Madani, K.: A Real-time Robot Vision Approach Combining Visual Saliency and Unsupervised Learning. In: Proc. of 14th Int. Conf. CLAWAR, Paris, France, pp. 241–248 (2011)
17. Ramik, D.-M., Sabourin, C., Madani, K.: Hybrid Salient Object Extraction Approach with Automatic Estimation of Visual Attention Scale. In: Proc. of Seventh Int. Conf. on Signal Image Technology & Internet-Based Systems, Dijon, France, pp. 438–445 (2011)
18. Ramik, D.M., Sabourin, C., Moreno, R., Madani, K.: A Machine Learning based Intelligent Vision System for Autonomous Object Detection and Recognition. J. of Applied Intelligence (2013), doi:10.1007/s10489-013-0461-5
19. Hofmann, J., Jüngel, M., Lötzsch, M.: A vision based system for goal-directed obstacle avoidance used in the rc'03 obstacle avoidance challenge. In: Proc. of 8th Int. Workshop on RoboCup, pp. 418–425 (2004)
20. Moreno, R., Ramik, D.M., Graña, M., Madani, K.: Image Segmentation on the Spherical Coordinate Representation of the RGB Color Space. IET Image Processing 6(9), 1275–1283 (2012)
21. Ramik, D.M., Sabourin, C., Madani, K.: Autonomous Knowledge Acquisition based on Artificial Curiosity: Application to Mobile Robots in Indoor Environment. J. of Robotics and Autonomous Systems 61(12), 1680–1695 (2013)

A Statistical Approach to Human-Like Visual Attention and Saliency Detection for Robot Vision: Application to Wildland Fires' Detection

Viachaslau Kachurka[1,2], Kurosh Madani[1], Christophe Sabourin[1], and Vladimir Golovko[2]

[1] Signals, Images, and Intelligent Systems Laboratory (LISSI / EA 3956), University Paris-Est Creteil, Senart-FB Institute of Technology, 36-37 rue Charpak, 77127 Lieusaint, France
[2] Neural Networks Laboratory, Intelligent Information Technologies Department, Brest State Techical University, Moskowskaya st. 267, 224017 Brest, Belarus
viachaslau.kachurka@gmail.com, {madani,sabourin}@u-pec.fr, gva@bstu.by

Abstract. In this work we contribute to development of a real-time human-like intuitive artificial vision system taking advantage from visual attention skill. Implemented on a 6-wheels mobile robot equipped with communication facilities, such a system allows detecting combustion perimeter in real outdoor environment without prior knowledge. It opens appealing perspectives in fire-fighting strategy enhancement and in early-stage woodland fire's detection. We provide experimental results showing as well the plausibility as the efficiency of the proposed system.

Keywords: Intuitive vision, saliency detection, fire region detection, robot, implementation.

1   Introduction

Wildland fires represent a major risk for many Mediterranean countries like France, Spain, Portugal and Greece but also other regions around the world such as Australia, California and recently Russia. They often result in significant human and economic losses. For example, in 2009 in Australia alone, 300 000 hectares were devastated and 200 people killed. The situation is so dramatic, that in the recent past (May 2007), the FAO (Food and Agriculture Organization of the United Nations) stressed the world governments to take actions for a better prevention, understanding and fighting of wildfires [1].

If the modeling of combustion and the prediction of fire front propagation remain foremost subjects in better prevention of fire disasters (be it wildland fire or compartment fire), efficient fire fighting lingers a curial need in precluding the dramatic consequences of such disasters. Such efficiency may be reached, or at least substantially enhanced, with information relative to the current state and the dynamic evolution of fires. For example, GPS systems make it possible to know the current

V. Golovko and A. Imada (Eds.): ICNNAI 2014, CCIS 440, pp. 124–135, 2014.

position of the resources and satellite images can be used to sense and track fire. However, the time scale and spatial resolution of these systems are still insufficient for the needs of operational forest fires fighting [2]. Fig. 1 shows views of wildland fires (forests' fires) and firefighting difficulty inherent to the intervention's conditions. From this sample of pictures it could be visible that if GPS and satellite images may help in locating and appreciation of overall fire, they remain inefficient in appreciation of local characteristics or persons to be rescued. On the other hand, over the two passed decades, visual and infrared cameras have been used as complementary metrological instruments in flame experiments [3]. Vision systems are now capable of reconstructing a 3D turbulent flame and its front structure when the flame is the only density field in images [4]. These 3D imaging systems are very interesting tools for flame study, however they are not suitable for large and wildland fires. In fact, the color of the fire depends on the experimental conditions. Outdoor fires are characterized by dominant colors in yellow, orange and red intervals. In outdoor experiments, the scenes are unstructured with inhomogeneous fuel of various colors (green, yellow, brown). Moreover overall light conditions may vary influencing the fires' aspect and their natural colors. Recent research in fire region segmentation has shown promising results ([5], [6], [7] and [8]). However, comparison tests showed that none of these techniques is well suited for the segmentation of all fire scenarios.

Fig. 1. Example showing firefighters intervention difficulty

Authors of [9] have recently developed a number of works based on the use of a stereovision system for the measurement of geometrical characteristics of a fire front during its propagation. Experiments were carried out in laboratory and in semi open field. The first part of this work corresponds to a segmentation procedure in order to extract the fire areas from the background of the stereoscopic images. However, although showing a number of promising results, especially relating the 3D aspects, as other referenced recent research works in fire region segmentation, the same work concluded that none of proposed techniques is well suited for the segmentation of all fire scenarios. So, the critical need (and the already open problem) of efficient extraction of only the region containing fire data remains entire. Moreover, in order to be useful for online fire-fighting strategy's updating the fire's segmentation and detection have to satisfy real-time computational requirements: be compatible with human's reactivity and thus not exceeding the scale-time of "second".

In contrast to the limitations of the aforementioned approaches, humans rely strikingly much on their natural vision in their perception of surrounding environment. By means of their "perceptual curiosity" and visual attention, humans can intuitively, reflexively and promptly detect and extract relevant items in diverse situations and within complex environments. Thus, it appears appropriate to draw inspiration from studies on human infants' learning-by-demonstration where perceptual curiosity stimulates children to discover what is salient to learn. Experiments in [10] show that it is the explicitness or exaggeration of an action that helps a child to understand, what is important in the actual context of learning. It may be generalized, that it is the saliency (motion, colors, etc.) that lets the pertinent information "stand-out" from the context and become "surprising" [11]. We argue that in this context the visual saliency may be helpful to enable unsupervised extraction of relevant items in complex or confusing environment as those in line with natural disasters including fire disasters.

Visual saliency (also referred in literature as visual attention, unpredictability or surprise) is described as a perceptual quality that makes a region of image stand out relative to its surroundings and to capture attention of observer [12]. The inspiration for the concept of visual saliency comes from the functioning of early processing stages of human vision system and is roughly based on previous clinical research. In early stages of the visual stimulus processing, human vision system first focuses in an unconscious, bottom-up manner, on visually attractive regions of the perceived image. The visual attractiveness may encompass features like intensity, contrast and motion. Although there exist biologically based approaches to visual saliency computation, most of the existing works do not claim to be biologically plausible. Instead, they use purely computational techniques to achieve the goal. One of the first works using visual saliency in image processing has been published by [13]. Authors use a biologically plausible approach based on a center-surround contrast calculation using Difference of Gaussians. Published more recently, other common techniques of visual saliency calculation include graph-based random walk [14], center-surround feature distances [15], multi-scale contrast, center-surround histogram and color spatial distribution or features of color and luminance [16]. A less common approach is described in [17]. It uses content-sensitive hyper-graph representation and partitioning instead of using more traditional fixed features and parameters for all images. Finally, in their recent works, authors of [18] and [19] have investigated an intelligent autonomous salient vision system for humanoid robots' autonomous knowledge acquisition as well as a powerful segmentation approach taking advantage from RGB representation in spherical coordinates [20].

In this paper, in view of the above-mentioned latest works, we present a fast visual saliency detection technique taking advantage from a statistical slant of human-like visual attention, with application to real-time fire detection and segmentation in natural outdoor environment. In Section 2, we detail our approach by outlining its principle. Section 3 focuses the implementation on a 6-wheels mobile robot (namely Wifibot-M from NEXTER Robotics) and the validation of the proposed approach in outdoor environment. Finally, Section 4 concludes the paper and outlines future work.

2   Visual Saliency and Visual Attention

Visual saliency is computed using saliency features, called Global Saliency Map and Local Saliency Map, respectively. These features play the role of some kind of saliency indicators relating luminance and chromaticity characterizing items of the image. The Global Saliency Map relates global variance of luminance and chromaticity of image while Local Saliency Map deals with locally centered (e.g. focusing a local region) characteristics within the image. The final saliency map is a nonlinear fusion of these two kinds of saliency features.

2.1   Global and Local Saliency Maps

Let us suppose the image Ω_{RGB} (acquired by the robot's visual sensor, namely its camera) represented by its pixels $\Omega_{RGB}(x)$ in RGB color space, where $x \in \aleph^2$ denotes its 2-D pixels' coordinates, and Ω_{YCC} the same image represented by its pixels $\Omega_{YCC}(x)$ in YC$_r$C$_b$ color space. Let $\Omega_R(x)$, $\Omega_G(x)$ and $\Omega_B(x)$ be the colors values in channels R, G and B, respectively. Let $\Omega_Y(x)$, $\Omega_{Cr}(x)$ and $\Omega_{Cb}(x)$ be the colors values in channels Y, C$_r$ and C$_b$, respectively. Finally, let Ω_Y^μ, Ω_{Cr}^μ and Ω_{Cb}^μ be median values computed for each channel (e.g. channels Y, C$_r$ and C$_b$) throughout the whole image.

The Global Saliency Map, denoted $M(x)$, is a result of nonlinear fusion of two elementary maps $M_Y(x)$ and $M_{CrCb}(x)$ relating luminance and chromaticity separately. The equation (1) details the calculation of each elementary map as well as the resulting Global Saliency Map.

$$M_Y(x) = \left\| \Omega_Y^\mu - \Omega_Y(x) \right\|$$

$$M_{CrCb}(x) = \sqrt{\left[\Omega_{Cr}^\mu - \Omega_{Cr}(x) \right]^2 + \left[\Omega_{Cb}^\mu - \Omega_{Cb}(x) \right]^2}$$

$$M(x) = \frac{1}{1 - e^{-C(x)}} M_{CrCb}(x) + \left(1 - \frac{1}{1 - e^{-C(x)}} \right) M_Y(x) \tag{1}$$

Coefficient C(x), determined using equation (2), is defined as value linking saturation C$_c$ of each pixel (note the C$_c$ is computed in RGB color space).

$$C(x) = 10(C_c(x) - 0.5),$$

$$C_c(x) = \max(\Omega_R(x), \Omega_G(x), \Omega_B(x)) - \min(\Omega_R(x), \Omega_G(x), \Omega_B(x)) \tag{2}$$

The purpose of local saliency meets the idea of center-surround difference of histograms (feature originally inspired by [21]). The idea is to go through the entire image and to compare the content of a sliding window with its surroundings to determine, how similar the two are. If the similarity is low, then it may be a sign of a salient region. Let us have a sliding window P of size p, centered over pixel (x).

Then let us define a (center) histogram H_C of pixel intensities inside it and a (surround) histogram H_S as histogram of intensities in a window Q surrounding P in a manner that the area of $(Q - P) = p^2$ (see Fig. 2-a). Then center-surround feature $d(x)$ is given as expressed in equation (3) over all histogram bins (i) (with $i \in \{Y, Cr, Cb\}$), where $|H_C|$ and $|H_S|$ are pixel counts for each histogram allowing it to be normalized although a part of the windows is out of the image frame. In this case only pixels inside the image are counted.

Window P   **Window Q**

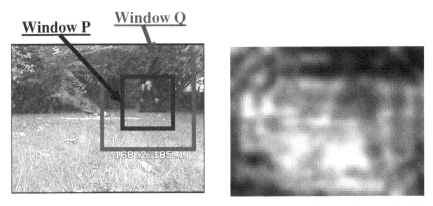

Fig. 2. The idea of center-surround difference of histograms (left-side) and example of local saliency map (right-side)

$$d_i(x) = \sum_{i=1}^{255}\left(\frac{H_C(i)}{|H_C|} - \frac{H_S(i)}{|H_S|} \right) \tag{3}$$

Calculating the $d(x)$ throughout all the Y, C_r and C_b channels, one can compute the resulting center-surround saliency D on a given position (x) as it is shown in equation (4), where Cμ is the average color saturation over the sliding window P. It is computed from RGB color model for each pixel as normalized ($0-1$ range) pseudo-norm accordingly to equation (5), where C is obtained similarly from equation (2). When C is low (too dull, unsaturated colors) importance is given to intensity saliency. When C is high (vivid colors) chromatic saliency is emphasized. Fig. 2-b gives an example of resulting so-called local saliency map.

$$D(x) = \frac{1}{1 - e^{-C_\mu(x)}} d_Y(x) + \left(1 - \frac{1}{1 - e^{-C(x)}}\right) Max(d_{Cr}(x), d_{Cb}(x)) \tag{4}$$

$$C_\mu(x) = \frac{\sum_{k \in P(x)}^{p} C(k)}{p} \tag{5}$$

Over size p (of the previously defined P sliding window), local features $D(x)$ are scale-dependent. Thus the parameter p may play the role of "visual attention" control parameter driving the saliency extraction either in the direction of roomy items' relevance (impact of high p value) or toward details' highlighting (impact of low p value). Such visual attention parameter allows a top-down control as well of the attention as of the sensitivity of the feature in scale space. High p value (resulting in a large sliding window size) with respect to the image size will make the local saliency feature more sensitive to large coherent parts of the image. On the other hand, low values of p will allow focusing to smaller details. For example, considering the human's face image, larger p will leads to extraction of the entire face, while lower p will focus on smaller items of the face as: eyes, lips, etc...

2.2   Segmentation

The purpose of segmentation is to match regions in image representing homogeneity (e.g. potential objects' shapes). While [21] used spherical representation of RGB color space, we use YCrCb color space although this representation remains more unkind then the previous one. The process is a threshold based weighted fusion operation resulting on a segmented regions map computed using a hybrid distance measure between two pixels in the image. Equation (6) details the computation of such a segmented regions map $d_{Hyb}(x, x')$, where $\alpha(x, x')$ and $\overline{\alpha}(x, x') = 1 - \alpha(x, x')$ stand for the fusion's weights with respect to the condition $\alpha(x, x') + \overline{\alpha}(x, x') = 1$. $d_Y(x, x')$), $d_{CrCb}(x, x')$ and $\alpha(d_{Hyb})$ are expressed by equations (7) and (8), respectively. The values of a, b and c are determined heuristically using different images in different conditions including different objects and shapes.

$$d_{Hyb}(x, x') = \alpha(x, x')d_Y(x, x') + \overline{\alpha}(x, x')d_{CrCb}(x, x') \tag{6}$$

$$d_Y(x, x') = \left\| \Omega_Y(x) - \Omega_Y(x') \right\| \tag{7}$$

$$d_{CrCb}(x, x') = \sqrt{\left[\Omega_{Cr}(x) - \Omega_{Cr}(x')\right]^2 + \left[\Omega_{Cb}(x) - \Omega_{Cb}(x')\right]^2}$$

$$\alpha(d_{Hyb}) = \begin{cases} 0 & \text{if } d_{Hyb} \le a \\ \dfrac{c}{2} + \dfrac{c}{2} Sin\left(\dfrac{\pi(d_{Hyb} - a)}{b - a} + \pi\right) & \text{if } a < d_{Hyb} < b \\ c & \text{if } d_{Hyb} \ge b \end{cases} \tag{8}$$

2.3   Salient Objects' Extraction

The saliency extraction is performed using two main features. The first one is the aforementioned segmented regions map $d_{Hyb}(x, x')$, detailed in previous subsection.

The second one is what we called "Final-Saliency-Map" (FSM), denoted by $SFM_{GB}(x)$. It is the result of a Gaussian-like blurring of a map resulting from the threshold-based fusion of global and local saliency maps (detailed in subsection 2.1). The parameter σ represents the radius of Gaussian smoothing function taking the value $\sigma = 5$ in our case. Equation (9) details the so called FSM, where $M_F(x)$ (equation (10)) represents the result involving global and local saliency maps' fusion.

$$SFM_{GB}(x) = GaussianBlur(M_F(x), \sigma) \qquad (9)$$

$$M_F(x) = \begin{cases} D(x) & if \ D(x) < M(x) \\ \sqrt{M(x) D(x)} & else \end{cases} \qquad (10)$$

Once $SFM_{GB}(x)$ available, for each homogeneous segment S_i from $d_{Hyb}(x,x')$ belonging to the ensemble of segments (e.g. potential objects in image) found in $d_{Hyb}(x,x')$ (e.g. $\forall S_i \in \{S_1, \cdots, S_i, \cdots, S_n\}$), the statistical central momentum $\mu(S_i)$ and variance $Var(S_i)$ of S_i are calculated from corresponding pixels of that segment in FSM (e.g. from $SFM_{GB}(x)$) accordingly to the operations expressed in equation (11). These two statistical momentums serve then to generate a binary map (e.g. a binary mask) filtering salient objects from the input image. The resulting map $M_{Mask}(x)$ is an image containing only salient objects where trifling objects have been replaced by black pixels. Equation (12) details this last operational step.

$$\mu(S_i) = \frac{\sum_{x \in S_i} SFM_{GB}(x)}{|S_i|} \qquad (11)$$

$$Var(S_i) = \sum_{x \in S_i} (SFM_{GB}(x) - \mu(S_i))^2$$

$$M_{Mask}(x) = \begin{cases} 1 & if \ \mu(S_i) > \tau_\mu \ and \ Var(S_i) > \tau_{Var} \\ 0 & else \end{cases} \qquad (12)$$

τ_μ and τ_{Var} represent matching threshold. The value of τ_μ corresponds to 50% of maximal saliency (e.g. $\forall x \in \aleph^2$ and $x \in S_i$: $\tau_\mu = 0.5 Max(SFM_{GB}(x))$). While the value of τ_{Var}, its magnitude has been set to 50% of nominal value of an ideal Gaussian distribution (e.g. $0.5 \times \left(1 - \frac{\sqrt{2}}{2}\right) = 0.5 \times 30\% = 15\%$). For example, if the maximum grey-level dynamics is 256, then $\tau_\mu = 128$ and $\tau_{Var} \cong 20$. Fig. 3 gives examples of different maps from input image to the extracted fire's shape comparing the extracted shape with the shape determined manually by confirmed professional firefighter expert. The input image (e.g. Fig.3-a) as well as the expert-based fire's shape determination (e.g. Fig.3-c) are provided by UMR CNRS 6134 SPE laboratory of University of Corsica (Corsica, France). Fig.4 illustrates fire's area detection (and different related maps) using the proposed technique and image from robot's camera.

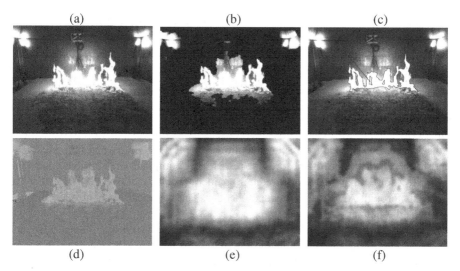

Fig. 3. Examples of fire's shape detection showing different maps resulting from saliency extraction process: input image (a), extracted fire's shape (b), fire's shape extracted by expert (c), segmented regions map (d), Global-Saliency-Map (f) and Local-Saliency-Map (f)

Fig. 4. Examples of different maps resulting from saliency extraction process: input image (a), extracted fire's shape (b), Final-Saliency-Map (c), segmented regions map (d) and the M_{Mask} map (e)

3   Implementation Using Real Robot and Validation Results

The designed system has been implemented on a Wifibot-M, a 6-wheels mobile robot (from NEXTER Robotics). Initially constructed for local mobile surveillance applications, Wifibot-M is composed by a six-wheel-driven waterproof (IP64) polycarbonate chassis controllable using WIFI. The chassis is composed by 3 parts linked with a 2 dimensional link. Wifibot-M robot can handle devices such as IP camera (MJPEG or MPEG) or any Ethernet sensors. A liteStation2 from UBNT router is the main CPU allowing data transfer, however, a 5Ghz router can be added. The robot used in the frame of the present research is equipped with an analogue PTZ (Pan-Tilt-Zoom, three degrees of freedom) camera (WONWOO WCM-101), attached to chassis through AXIS M7001 video encoder. The robot can be controlled through WIFI or Ethernet network connection.

A software controller is shipped with Wifibot-M. Designed by robot producents, this controller is programmed in C++ with usage of WinAPI and DirectConnect technologies, making it a "Windows-dedicated" software. This controller can be run on any Windows PC connecting the same network as the robot. It allows the usage of any Plug-and-Play device compatible with Windows. Thus, it makes the robot to be controlled by any Plug-and-Play joystick, or its virtual simulation to control either the robot's movement or to collect data from robot's internal sensors, as those relating robot's speed or odometers' data from its wheels: this is somehow an interesting point regarding the aimed application involving human operator. Also it can switch camera on/off, while it cannot give the image from this camera to user. The image can be collected, using either Web-interface of AXIS Encoder or AXIS Camera Client application, which is also Windows-only software. Although the above-mentioned already available facilities, the last point remains a drawback for the focused application, because the involvement of an operator based guidance results on the necessity of combining both movement controller and video stream controller in a same unit. In fact the exploitation has to be lightweight for the user. Due to these conditions, a new controller has been designed, "Wifibot-M iPy Controller" allowing:

- Connection via network, switch on/off its camera and/or controlling its wheels;
- To control robot's movement by a Plug-and-Play joystick;
- To collect video stream from its camera and store frames as JPEG images;
- To control camera's PTZ-routines;
- To detect salient objects from video stream in human-compatible real-time;
- To react consequently (thus in human-compatible real-time) on collected results.

From an architectural slant of view, the designed controller could be seen as a 4-modules unit: movement controller, PTZ & video stream handler, salient regions detector and reaction strategies inspector. Fig. 5 depicts the general bloc diagram of such architecture. The main tasks-managing module includes a Graphical User Interface (GUI) and a number of standard task-handling routines. That is why this part is not shown in the Fig.5.

The validation scenario has been based on a real outdoor fire's detection situation. In order to make the validation scenario compatible with plausible fire-fighting circumstances, the scenario has been realized within the robot's scale. Accordingly to the Wifibot-M robot's size, this means some 300 m² area (typically $25 \times 12 m^2$ action-area) and a 80-to-100 centimeters-height and 100 centimeters-width fire with smoke somewhere in that area. This also ensures a correct WIFI connection (as well regarding network connection's quality as regarding the relative simplicity of required supply deployment in outdoor conditions) and a correct energetic autonomy of the robot allowing performing several experimental tests during several hours if required.

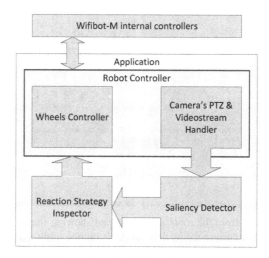

Fig. 5. Block diagram showing the implementation's architecture

Fig. 6. Experimental area showing the robot with its camera and the combustion zone

Several tests supposing robot moving toward and around the combustion (fire) zone with aim of detection of the fire's shape as salient target have been realized. Fig 6 depicts the experimental setup and the fire's perimeter. Fig. 7 gives the obtained results showing on the left-side pictures the robot's camera view of the scenery and on the right-side pictures the detected salient combustion area. As it is visible from right-side pictures of Fig. 7, the salient combustion zone is correctly detected. Moreover, it is pertinent to note that as well fire's outline as the smoke's perimeter have been correctly detected and recognized as salient objects and events. The detection of the smoky perimeter as salient item of the scenery is an interesting point because often the woodland fires generate smoky atmosphere which may be used as an early-stage salient indicator in early-detection of upcoming woodland fire deserter.

Fig. 7. Pictures extracted from the robot's video stream showing the robot's view of scenery (left-side pictures) and the corresponding detected fire's outline (right-side pictures)

4   Conclusion and Further Work

In this paper, we have presented and validated a real-time intuitive artificial vision system inspired from human-like vision. Taking advantage from visual attention capabilities, such a system may play a substantial role in enhancing fire-fighting in the context of woodland fire disasters. Implemented on a 6-wheels mobile robot equipped with vision and communication means, the investigated system shows as well the feasibility as effectiveness of the proposed solution. Further step will consist on enhancing the saliency detection for fire and smoke separately.

Aknoledgements. This work is performed in the frame of a close and continuous collaboration with UMR CNRS 6134 SPE lab. of University of Corsica (France). Authors wish to thank Dr. Lucile Rossi from UMR CNRS 6134 SPE for her help and useful discussions.

References

1. FAO, Wildfire management, a burning issue for livelihoods and land-use (2007), http://www.fao.org/newsroom/en/news/2007/1000570/index.html
2. San-Miguel-Ayanz, J., Ravail, N., Kelha, V., Ollero, A.: Active fire detection for emergency management: potential and limitations for the operational use of remote sensing. Natural Hazards 35, 361–376 (2005)

3. Lu, G., Yan, Y., Huang, Y., Reed, A.: An Intelligent Monitoring and Control System of Combustion Flames. Meas. Control 32(7), 164–168 (1999)

4. Gilabert, G., Lu, G., Yan, Y.: Three-Dimensional Tomographic Renconstruction of the Luminosity Distribution of a Combustion Flame. IEEE Trans. on Instr. and Measure. 56(4), 1300–1306 (2007)

5. Rossi, L., Akhloufi, M., Tison, Y.: Dynamic fire 3D modeling using a real-time stereovision system. J. of Communication and Computer 6(10), 54–61 (2009)

6. Ko, B.C., Cheong, K.H., Nam, J.Y.: Fire detection based on vision sensor and support vector machines. Fire Safety J. 44, 322–329 (2009)

7. Celik, T., Demirel, H.: Fire detection in video sequences using a generic color model. Fire Safety J. 44, 147–158 (2009)

8. Chen, T., Wu, P., Chiou, Y.: An early fire-detection method based on image processing. In: Proc. of Int. Conf. on Image Processing, pp. 1707–1710 (2004)

9. Rossi, L., Akhloufi, M., Tison, Y., Pieri, A.: On the use of stereovision to develop a novel instrumentation system to extract geometric fire fronts characteristics. Fire Safety Journal 46(1-2), 9–20 (2011)

10. Brand, R.J., Baldwin, D.A., Ashburn, L.A.: Evidence for 'motionese': modifications in mothers infant-directed action. Developmental Science, 72–83 (2002)

11. Wolfe, J.M., Horowitz, T.S.: What attributes guide the deployment of visual attention and how do they do it? Nature Reviews Neuroscience, 495–501 (2004)

12. Achanta, R., Hemami, S., Estrada, F., Susstrunk, S.: Frequency-tuned Salient Region Detection. In: Proc. of IEEE Int. Conf. on Computer Vision and Pattern Recognition (2009)

13. Itti, L., Koch, C., Niebur, E.: A Model of Saliency-Based Visual Attention for Rapid Scene Analysis. IEEE Trans. on Pattern Analysis and Machine Intel. 20, 1254–1259 (1998)

14. Harel, J., Koch, C., Perona, P.: Graph-based visual saliency. In: Advances in Neural Information Processing Systems, vol. 19, pp. 545–552 (2007)

15. Achanta, R., Estrada, F., Wils, P., Süsstrunk, S.: Salient Region Detection and Segmentation. In: Gasteratos, A., Vincze, M., Tsotsos, J.K. (eds.) ICVS 2008. LNCS, vol. 5008, pp. 66–75. Springer, Heidelberg (2008)

16. Liu, T., Yuan, Z., Sun, J., Wang, J., Zheng, N., Tang, X., Shum, H.-Y.: Learning to Detect a Salient Object. IEEE Trans. Pattern Anal. Mach. Intell. 33(2), 353–367 (2011)

17. Liang, Z., Chi, Z., Fu, H., Feng, D.: Salient object detection using content-sensitive hypergraph representation and partitioning. Pattern Rec. 45(11), 3886–3901 (2012)

18. Ramík, D.M., Sabourin, C., Madani, K.: Hybrid Salient Object Extraction Approach with Automatic Estimation of Visual Attention Scale. In: Proc. of 7th Int. Conf. on Signal Image Technology & Internet-Based Systems, Dijon, France, pp. 438–445 (2011)

19. Ramik, D.M., Sabourin, C., Moreno, R., Madani, K.: A Machine Learning based Intelligent Vision System for Autonomous Object Detection and Recognition. J. of Applied Intelligence (2013), doi:10.1007/s10489-013-0461-5

20. Moreno, R., Ramik, D.M., Graña, M., Madani, K.: Image Segmentation on the Spherical Coordinate Representation of the RGB Color Space. IET Image Processing 6(9), 1275–1283 (2012)

21. Liu, T., Yuan, Z., Sun, J., Wang, J., Zheng, N., Tang, X., Shum, H.-Y.: Learning to Detect a Salient Object. In: Proc. of Computer Vision and Pattern Recognition, pp. 353–367 (2011)

A Learning Technique for Deep Belief Neural Networks

Vladimir Golovko[1], Aliaksandr Kroshchanka[1], Uladzimir Rubanau[1],
and Stanisław Jankowski[2]

[1] Brest State Technical University
Moskowskaja 267, Brest 224017, Belarus
gva@bstu.by
[2] Warsaw University of Technology
Nowowiejska 15/19, Warsaw 00-665, Poland

Abstract. Deep belief neural network represents many-layered perceptron and permits to overcome some limitations of conventional multilayer perceptron due to deep architecture. The supervised training algorithm is not effective for deep belief neural network and therefore in many studies was proposed new learning procedure for deep neural networks. It consists of two stages. The first one is unsupervised learning using layer by layer approach, which is intended for initialization of parameters (pretraining of deep belief neural network). The second is supervised training in order to provide fine tuning of whole neural network. In this work we propose the training approach for restricted Boltzmann machine, which is based on minimization of reconstruction square error. The main contribution of this paper is new interpretation of training rules for restricted Boltzmann machine. It is shown that traditional approach for restricted Boltzmann machine training is particular case of proposed technique. We demonstrate the efficiency of proposed approach using deep nonlinear auto-encoder.

1 Introduction

Neural techniques have been successfully applied to many problems in different domains. The deep belief neural networks (DBNN) [1–4] have been found to have a better performance and more representational power than traditional neural networks. Due to the increasing interest in these kinds of models, this paper focuses on training paradigm for such neural networks. The DBNN consists of many hidden layers and can perform a deep hierarchical representation of the input data. So, the first layer could extract low-level features, the second layer could extract higher level features and so on [5]. Let us consider the related works in this domain. These kind of neural networks were investigated in many studies [1–7]. There exist two main techniques for the DBNN training (Fig.1).

The first one is based on restricted Boltzmann machine (RBM) [1, 2] and uses two steps for DBNN training. At the beginning the greedy layer-wise training procedure is used, namely first layer is trained and its parameters are fixed, after

V. Golovko and A. Imada (Eds.): ICNNAI 2014, CCIS 440, pp. 136–146, 2014.
© Springer International Publishing Switzerland 2014

this the next layer is trained etc. As a result the initialization of neural network is performed and we could use supervised learning for fine tuning parameters of whole neural networks.

The second technique for DBNN training uses auto-encoder approach for pre-training of each layer and after this fine tuned method is applied in supervised manner. In this case we train in the beginning the first layer as autoassociative neural network in order to minimize the reconstruction error. Then the hidden units are used as the input for the next layer and next layer is trained as auto-encoder. Finally fine-tuning all of parameters of neural network by supervised way is performed.

In this work we propose a new interpretation of learning rules for RBM. The conventional approach to training the RBM uses energy-based model. The proposed approach is based on minimization of reconstruction mean square error, which we can obtain using a simple iterations of Gibbs sampling. We have shown that classical equations for DBNN training are particular case of proposed technique.

The rest of the paper is organized as follows. Section 2 describes the conventional approach for restricted Boltzmann machine training based on energy model. In Sect. 3 we propose novel approach for inference of RBM training rules. Section 4 demonstrates the results of experiments and finally Sect. 5 gives conclusion.

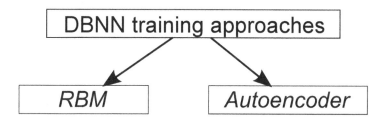

Fig. 1. DBNN training approaches

2   Deep Belief Neural Network Based on Restricted Boltzmann Machine

In this section we present a short introduction to deep belief neural networks based on restricted Boltzmann machine. The RBM is the main building block for deep belief neural networks. The conventional approach to RBM training is based on energy model and training rules take into account only linear nature of neural units.

As already mentioned, the DBNN consists of many hidden layers and can perform a deep hierarchical representation of the input data as shown in Fig. 2.

The j-th output unit for k-th layer is given by

$$y_j^k = F\left(S_j^k\right),\tag{1}$$

$$S_j^k = \sum_{i=1} w_{ij}^k y_i^{k-1} + T_j^k,\tag{2}$$

where F is the activation function, S_j^k is the weighted sum of the j-th unit, w_{ij}^k is the weight from the i-th unit of the $(k-1)$-th layer to the j-th unit of the k-th layer, T_j^k is the threshold of the j-th unit.

For the first layer

$$y_i^0 = x_i .\tag{3}$$

In common case we can write, that

$$Y^k = F\left(S^k\right) = F\left(W^k Y^{k-1} + T^k\right),\tag{4}$$

where W is a weight matrix, Y^{k-1} is the output vector for $(k-1)$-th layer, T^k is the threshold vector.

It should be also noted that the output of the DBNN is often defined using softmax function:

$$y_j^F = \text{softmax } S_j = \frac{e^{S_j}}{\sum_l e^{S_l}} .\tag{5}$$

Let's examine the restricted Boltzmann machine, which consists of two layers: visible and hidden (Fig. 3). The restricted Boltzmann machine is the part of the deep belief neural network and can represent any discrete distribution, if enough hidden units are used [5].

The layers of neural units are connected by bidirectional weights $\boldsymbol{W}$. In mostly cases the binary units are used [1–3]. The RBM is a stochastic neural network and the states of visible and hidden units are defined using probabilistic version of sigmoid activation function:

$$p\left(y_j|x\right) = \frac{1}{1 + e^{-S_j}},\tag{6}$$

$$p\left(x_i|y\right) = \frac{1}{1 + e^{-S_i}} .\tag{7}$$

The energy function of the binary state (x, y) is defined as

$$E\left(x, y\right) = -\sum_i x_i T_i - \sum_j y_j T_j - \sum_{i,j} x_i y_j w_{ij} .\tag{8}$$

The learning procedure of the RBM consists of presenting training patterns to the visible units and the goal is to update the weights and biases in order to minimize the energy of the network. A probability distribution through an energy function can be defined [5] as follows:

$$P(x) = \frac{1}{Z} \sum_y e^{-E(x,y)},\tag{9}$$

$$Z = \sum_{x,y} e^{-E(x,y)}, \tag{10}$$

where Z is the normalizing factor.

The gradient of the log-likelihood can be written as follows:

$$\frac{\partial \log P(x)}{\partial w_{ij}} = \langle x_i y_j \rangle_d - \langle x_i y_j \rangle_r, \tag{11}$$

$$\frac{\partial \log P(x)}{\partial T_i} = \langle x_i \rangle_d - \langle x_i \rangle_r, \tag{12}$$

$$\frac{\partial \log P(x)}{\partial T_j} = \langle y_j \rangle_d - \langle y_j \rangle_r. \tag{13}$$

As a result we can obtain the RBM training rules [2] as follows:

$$w_{ij}(t+1) = w_{ij}(t) + \alpha \left(\langle x_i y_j \rangle_d - \langle x_i y_j \rangle_r \right), \tag{14}$$

$$T_i(t+1) = T_i(t) + \alpha \left(\langle x_i \rangle_d - \langle x_i \rangle_r \right), \tag{15}$$

$$T_j(t+1) = T_j(t) + \alpha \left(\langle y_j \rangle_d - \langle y_j \rangle_r \right). \tag{16}$$

Here α is learning rate, $<>_d$ denotes the expectation for the data distribution and $<>_r$ denotes the expectation for the reconstructed (model) data distribution. Since computing the expectation using RBM model is very difficult, Hinton propose to use an approximation of that term called contrastive divergence (CD) [2]. It is based on Gibbs sampling. In this case the first term in equation (11) denotes the data distribution at the time $t = 0$ and the second term is distribution of reconstructed states at the step $t = n$. Therefore the CD-n procedure can be represented as follows:

$$x_0 \rightarrow y_0 \rightarrow x_1 \rightarrow y_1 \rightarrow \ldots \rightarrow x_n \rightarrow y_n. \tag{17}$$

Training an RBM is based on presenting a training sample to the visible units, then using the CD-n procedure we could compute the binary states of the hidden units $p(y|x)$, perform sampling the visible units (reconstructed states) $p(x|y)$ and so on. After performing these iterations the weights and biases of restricted Boltzmann machine are updated. Then we should stack another hidden layer to train a new RBM. This approach is applied to all layer of deep belief neural network (greedy layer-wise training). As a result of such an unsupervised pre-training we can obtain a good initialization of the neural network. Finally supervised fine-tuning algorithm of whole neural network is performed.

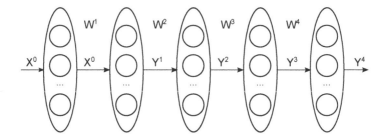

Fig. 2. Deep belief neural network structure

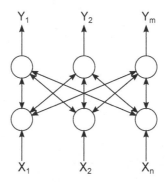

Fig. 3. Restricted Boltzmann machine

3   Alternative Approach for RBM Training Rules

In this section we propose a novel approach in order to infer RBM training rules. It is based on minimization of reconstruction mean square error, which we can obtain using a simple iterations of Gibbs sampling. In comparison with traditional energy-based method, which is based on linear representation of neural units, the proposed approach permits to take into account nonlinear nature of neural units.

Let's examine the restricted Boltzmann machine. We will represent the RBM, using three layers (visible, hidden and visible) [8] as shown in Fig. 4.

Let $x(0)$ will be input data, which move to the visible layer at time 0. Then the output of hidden layer is defined as follows:

$$y_j(0) = F\left(S_j(0)\right),\tag{18}$$

$$S_j(0) = \sum_i w_{ij} x_i(0) + T_j \ .\tag{19}$$

The inverse layer reconstructs the data from hidden layer. As a result we can obtain $x(1)$ at time 1:

$$x_i(1) = F(S_i(1)), \tag{20}$$

$$S_i(1) = \sum_j w_{ij} y_j(0) + T_i . \tag{21}$$

After this the $x(1)$ enters to the visible layer and we can obtain the output of the hidden layer by the following way:

$$y_j(1) = F(S_j(1)), \tag{22}$$

$$S_j(1) = \sum_i w_{ij} x_i(1) + T_j . \tag{23}$$

The purpose of the training this neural network is to minimize the reconstruction mean squared error (MSE):

$$E_s = \frac{1}{2} \sum_{k=1}^{L} \sum_{i=1}^{n} \left(x_i^k(1) - x_i^k(0)\right)^2 + \frac{1}{2} \sum_{k=1}^{L} \sum_{j=1}^{m} \left(y_j^k(1) - y_j^k(0)\right)^2, \tag{24}$$

where L is the number of training patterns. If we use online training of RBM the weights and thresholds are updated iteratively in accordance with the following rules:

$$w_{ij}(t+1) = w_{ij}(t) - \alpha \frac{\partial E}{\partial w_{ij}(t)}, \tag{25}$$

$$T_i(t+1) = T_i(t) - \alpha \frac{\partial E}{\partial T_i(t)}, \tag{26}$$

$$T_j(t+1) = T_j(t) - \alpha \frac{\partial E}{\partial T_j(t)} . \tag{27}$$

The cost function E for one sample is defined by the expression:

$$E = \frac{1}{2} \sum_i (x_i(1) - x_i(0))^2 + \frac{1}{2} \sum_j (y_j(1) - y_j(0))^2 . \tag{28}$$

Differentiating (28) with respect to w_{ij}, T_i and T_j we can get the following training rules for RBM network:

$$w_{ij}(t+1) = w_{ij}(t) - \alpha \left((x_i(1) - x_i(0)) F'(S_i(1)) y_j(0) + \right.$$
$$\left. + (y_j(1) - y_j(0)) F'(S_j(1)) x_i(1)\right), \tag{29}$$

$$T_i(t+1) = T_i(t) - \alpha (x_i(1) - x_i(0)) F'(S_i(1)), \tag{30}$$

$$T_j(t+1) = T_j(t) - \alpha (y_j(1) - y_j(0)) F'(S_j(1)) . \tag{31}$$

We can use these rules for training RBM network for any data (binary and real). Let's examine the interrelation between conventional and proposed RBM training rules. Let us suppose, that linear activation function is used. This is equivalent to

$$F'\left(S_i(1)\right) = \frac{\partial x_i(1)}{\partial S_i(1)} = 1 \text{ and } F'\left(S_j(1)\right) = \frac{\partial y_j(1)}{\partial S_j(1)} = 1 \ . \tag{32}$$

Then the training rules can transform by the following way:

$$w_{ij}(t+1) = w_{ij}(t) + \alpha\left(x_i(0)y_j(0) - x_i(1)y_j(1)\right), \tag{33}$$

$$T_i(t+1) = T_i(t) + \alpha\left(x_i(0) - x_i(1)\right), \tag{34}$$

$$T_j(t+1) = T_j(t) + \alpha\left(y_j(0) - y_j(1)\right) \ . \tag{35}$$

As we can see the last equations are identical to the conventional RBM training rules. Thus the classical equations for RBM training are particular case of proposed technique.

If we use the CD-n procedure

$$w_{ij}(t+1) = w_{ij}(t) - \alpha\left((x_i(n) - x_i(0))\, F'\left(S_i(n)\right) y_j(n-1) + \right.$$
$$\left. + (y_j(n) - y_j(0))\, F'\left(S_j(n)\right) x_i(n)\right), \tag{36}$$

$$T_i(t+1) = T_i(t) - \alpha\left(x_i(n) - x_i(0)\right) F'\left(S_i(n)\right), \tag{37}$$

$$T_j(t+1) = T_j(t) - \alpha\left(y_j(n) - y_j(0)\right) F'\left(S_j(n)\right) \ . \tag{38}$$

If $x(n) = x(1)$, then $y(n-1) = y(0)$ and we can get the conventional RBM equations:

$$w_{ij}(t+1) = w_{ij}(t) + \alpha\left(x_i(0)y_j(0) - x_i(n)y_j(n)\right), \tag{39}$$

$$T_i(t+1) = T_i(t) + \alpha\left(x_i(0) - x_i(n)\right), \tag{40}$$

$$T_j(t+1) = T_j(t) + \alpha\left(y_j(0) - y_j(n)\right) \ . \tag{41}$$

If the batch learning is used we can write:

$$w_{ij}(t+1) = w_{ij}(t) - \alpha\frac{\partial E_s}{\partial w_{ij}(t)}, \tag{42}$$

$$T_i(t+1) = T_i(t) - \alpha\frac{\partial E_s}{\partial T_i(t)}, \tag{43}$$

$$T_j(t+1) = T_j(t) - \alpha\frac{\partial E_s}{\partial T_j(t)} \ . \tag{44}$$

Then we can obtain the following equations for RBM training using CD-n procedure:

$$w_{ij}(t+1) = w_{ij}(t) - \alpha \sum_{k=1}^{L} \left(\left(x_i^k(n) - x_i^k(0) \right) F' \left(S_i^k(n) \right) y_j^k(n-1) + \right.$$
$$\left. + \left(y_j^k(n) - y_j^k(0) \right) F' \left(S_j^k(n) \right) x_i^k(n) \right), \quad (45)$$

$$T_i(t+1) = T_i(t) - \alpha \sum_{k=1}^{L} \left(x_i^k(n) - x_i^k(0) \right) F' \left(S_i^k(n) \right), \quad (46)$$

$$T_j(t+1) = T_j(t) - \alpha \sum_{k=1}^{L} \left(y_j^k(n) - y_j^k(0) \right) F' \left(S_j^k(n) \right). \quad (47)$$

In this section we have obtained the training rules for restricted Boltzmann machine. It is based on minimization of reconstruction mean square error, which we can obtain using a simple iterations of Gibbs sampling. The proposed approach permits to take into account the derivatives of nonlinear activation function for neural network units. We will call the proposed approach reconstruction error-based approach (REBA). It was shown, that the classical equations for RBM training are particular case of proposed technique.

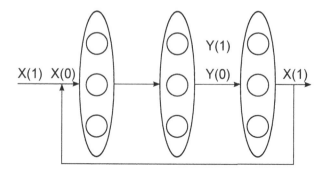

Fig. 4. Expanded representation of RBM

4   Experimental Results

To assess the performance of the proposed learning technique experiments were conducted on artificial data set. The artificial data x lie on a one-dimensional manifold (a helical loop) embedded in three dimensions [9] were generated from a uniformly distributed factor t in the range [-1,1]:

$$\begin{cases} x_1 = \sin(\pi t) \\ x_2 = \cos(\pi t) \\ x_3 = t \end{cases} \quad (48)$$

In order to verify the proposed approach experimentally, we train seven-layer auto-encoder deep belief neural network using data subsets of 1000 samples. The deep auto-encoder is shown in Fig. 5. We used the sigmoid activation function for all layers of neural network except for bottleneck layer. The linear activation function is used in bottleneck layer (fourth layer).

The average results are provided in the Table 1. Here NIT is the number of training epochs, using fine-tuning algorithm for whole neural network; MSE – mean square error on the training data set, MS – mean square error on the test data set in order to check generalization ability. The size of test patterns is 1000. As can be seen, the training procedure works well if Gibbs sampling is used only for one iteration (CD-1). The learning rate α is 0.01 for all experiments.

Table 1. Comparison of RBM and REBA techniques

Training procedure	n for CD-n	MSE	MS	NIT
RBM	1	0,587	0,580	200
	5	1,314	1,290	200
	10	0,503	0,533	198
	20	0,477	0,488	200
REBA	1	0,442	0,475	199
	5	0,607	0,609	200
	10	0,870	0,875	200
	20	1,495	1,505	200

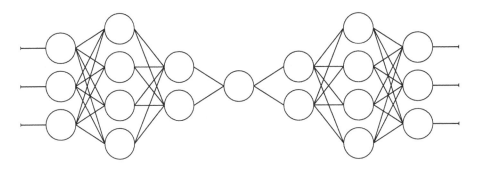

Fig. 5. Auto-encoder deep belief neural network

Figure 6 depicts the evolution of mean square error depending on epochs for the first layer of deep auto-encoder. The number of epochs for training of each layer is 50.

It is evident from the simulation results that the use of the REBA technique can improve the generalization capability of deep auto-encoder in case of CD-1 and CD-5.

Figure 7 shows the original training data and the reconstructed data from one component, using test data. As can be seen, the auto-encoder reconstructs the data from one nonlinear component with well accuracy.

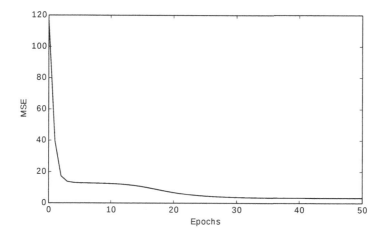

Fig. 6. Evolution of MSE on the first layer

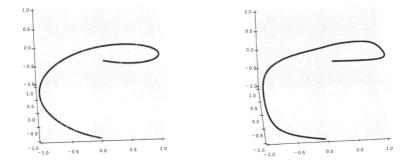

Fig. 7. 3D views of original and reconstructed datasets

5   Conclusion

In this paper we have addressed some key aspects of deep belief neural network training. We described both traditional energy-based method, which is based on linear representation of neural units and proposed approach, which is based on nonlinear neurons. The proposed approach is based on minimization of reconstruction mean square error, which we can obtain using a simple iterations of Gibbs sampling. As can be seen, the classical equations for RBM training are particular case of proposed technique. The simulation results demonstrate an efficiency of proposed technique.

References

1. Hinton, G.E., Osindero, S., Teh, Y.: A fast learning algorithm for deep belief nets. Neural Computation 18, 1527–1554 (2006)
2. Hinton, G.: Training products of experts by minimizing contrastive divergence. Neural Computation 14, 1771–1800 (2002)
3. Hinton, G., Salakhutdinov, R.: Reducing the dimensionality of data with neural networks. Science 313(5786), 504–507 (2006)
4. Hinton, G.E.: A practical guide to training restricted Boltzmann machines (Tech. Rep. 2010–000). Machine Learning Group, University of Toronto, Toronto (2010)
5. Bengio, Y.: Learning deep architectures for AI. Foundations and Trends in Machine Learning 2(1), 1–127 (2009)
6. Bengio, Y., Lamblin, P., Popovici, D., Larochelle, H.: Greedy layer-wise training of deep networks. In: Schölkopf, B., Platt, J.C., Hoffman, T. (eds.) Advances in Neural Information Processing Systems, vol. 11, pp. 153–160. MIT Press, Cambridge (2007)
7. Erhan, D., Bengio, Y., Courville, A., Manzagol, P.-A., Vincent, P., Bengio, S.: Why does unsupervised pre-training help deep learning? Journal of Machine Learning Research 11, 625–660 (2010)
8. Golovko, V., Vaitsekhovich, H., Apanel, E., Mastykin, A.: Neural network model for transient ischemic attacks diagnostics. Optical Memory and Neural Networks (Information Optics) 21(3), 166–176 (2012)
9. Scholz, M., Fraunholz, M., Selbig, J.: Nonlinear principal component analysis: neural network models and applications. In: Principal Manifolds for Data Visualization and Dimension Reduction, pp. 44–67. Springer, Heidelberg (2008)

Modeling Engineering-Geological Layers by k-nn and Neural Networks

Stanisław Jankowski[1], Alesia Hrechka[2], Zbigniew Szymański[1], and Grzegorz Ryżyński[3]

[1] Warsaw University of Technology, Warsaw, Poland
sjank@ise.pw.edu.pl, z.szymanski@ii.pw.edu.pl
[2] Brest State Technical University, Brest, Belarus
AlesiaHrechka@gmail.com
[3] Polish Geological Institute – National Research Institute, Warsaw, Poland
grzegorz.ryzynski@pgi.gov.pl

Abstract . In this paper a novel approach for solving task of engineering geological layers approximation is described. The task is confined in performing smooth geological surface from the heterogenic array of points in 3-dimensional space. This approach is based on statistical learning systems, as: k-nn (k-nearest neighbors) algorithm and neural networks (multilayer perceptron) which allows to separate points belonging to different geological layers. Our method enables also modeling convex 3-dimensional intrusions. The main advantage of our approach is the possibility of the surface model scaling without increasing the calculation complexity.

Keywords: k-nn classifier, neural network approximation, engineering-geological layer, geological mapping, geological cartography.

1   Introduction

The online presentation of geological data and information, fostered by the development of international spatial data infrastructure, demands an effective numerical algorithms for visualization and analysis of geological structures.

The public institutions in different countries, the National Geological Surveys, are collecting and managing the very large datasets of geological boreholes (exceeding tens thousands of boreholes), which are then the basis for preparation of geological maps, cross-sections and geological models (2D, 3D, and 4D). One of the goals of this paper was to assess the practical potential of statistical learning systems as an effective numerical tools for large geological datasets. The two methods presented in the paper are: k-nn (k nearest neighbours) and neural network approximation [1-7]. Both methods are characterized by the elastic adaptation to the natural complexity of the problem, also they have a vast areas of practical application. The paper is an attempt to answer if k-nn and neural networks approximation methods can be useful for interpolation of roof surfaces of engineering-geological layers. The potential areas of application were discussed (eg. lightweight mobile applications or online geo-reporting GIS systems).

V. Golovko and A. Imada (Eds.): ICNNAI 2014, CCIS 440, pp. 147–158, 2014.
© Springer International Publishing Switzerland 2014

The use of neural networks algorithms for construction of geological models was already discussed by many authors [8, 11, 12]. The presented paper is an attempt to continue research in this topic by application of statistical learning systems to a geological dataset of 1083 boreholes from the Engineering-Geological Atlas of Warsaw database managed by the Polish Geological Survey.

There are two main methods of engineering-geological layer modeling:

- Function approximation [8] f: $R^2 \rightarrow R$ – the depth of the layer roof is calculated using 2 arguments (point coordinates). One fundamental drawback is that complex geometry of the layer (e.g. containing folds) causes the method to fail. Advantages are its simplicity and short computation time.
- Volumetric approach based on classification of the voxels. There are two drawbacks of the volumetric method. It requires a lot of resources (e.g. memory, computation time). Some post processing is needed for visualization of such data. However this method offers modeling of layers of arbitrary shapes.

In this paper we present two methods of volumetric based approach: k- nearest neighbor classifier and neural network classifier.

2   Analyzed Dataset

The analyzed dataset are geological boreholes from Engineering-Geological Atlas of Warsaw Database. Full database is more than 27 000 boreholes. For the purpose of the presented paper the area of 3,5 x 4,0 km in the center of Warsaw was selected (see fig. 1). The selected subset consists of 1083 boreholes, of various depth, ranging from 1,5 to88,8 m below the ground level surface (m. b.g.l.s.). The average depth of boreholes is 14,3 m. The analyzed boreholes are documenting quaternary sediments (mostly antropogenic made ground of various composition, glacial tills and fluvioglacial sands and gravels) and Neogene sediments – the Pliocene clays. The boreholes in analyzed dataset were reclassified into 30 engineering-geological layers.

Each borehole in the dataset has a description of its lithological profile and its stratigraphy. On the basis of the lithology, stratigraphy and laboratory tests of physical and mechanical properties of collected soil samples the profile of each borehole is reclassified by engineering-geologists. This is made to define the engineering-geological layers of soils and rocks of similiar physico-mechanical properties (see fig. 3). Engineering-geological profiles of boreholes are necessary for evaluation of bearing capacity or for determination of engineering conditions for planed infrastructure (buildings, pipelines, highways, underground structures, etc.).

Fig. 1. The analyzed dataset consists of 1083 geological boreholes from an area of 3,5 x 4,0 km, located in the center of Warsaw

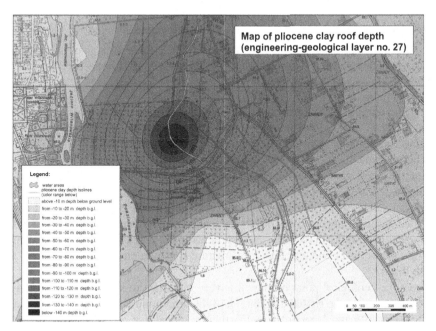

Fig. 2. Example of classically generated map of pliocene clay roof, using ordinary krigging method (from Polish Geological Institute archival materials)

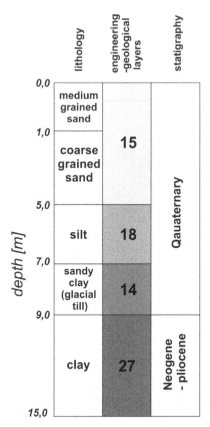

Fig. 3. Engineering-geological boreholes. Each borehole has a description of the lithological profile, stratigraphy and it is subdivided into engineering-geological layers (eg. layer no. 15 – fluvioglacial sands, layer no. 27. pliocene clays).

For the purpose of this paper the analysis of the geological data described above was focused mostly on engineering geological-layers of fluvioglacial sands (engineering geological layer no. 15) and pliocene clays (layer no. 27).

The pliocene clays on the area of Warsaw were highly glacitectonicaly deformed during the latest glacial period, their roof surface morphology is very undulated. In the depressed areas of pliocene clays roof surface there are often accumulated layers of glacial sands, which are saturated with the pressurized groundwater, that cannot infiltrate through impermeable layers of underlying pliocene clays. This kind of geological conditions in Warsaw is very unfavorable and can cause severe dangers during earthworks or underground infrastructure construction (eg. construction of new metro lines).

From the geological point of view it is very important to know the morphology of pliocene clays and the spatial distribution of the discontinuous lenses of fluvio-glacial sands to properly and accurately identify the geological risk.

The classical approach for visualisation of engineering-geological layers roof surface morphology is preparation of isoline maps with the use of ordinary kriging method or Euclidean allocation. The example of such map, made by kriging method is presented in the figure 2.

Table 1. Example data used for approximation. The dataset contained 10466 records from 1083 boreholes.

Borehole name	roof of layer [m, depth]	X coordinate [m] local coordinate system	Y coordinate [m] local coordinate system	(ground level) [m] local coordinate	Symbol	Engineering -geological layer	Description
O-2/226	0,00	-1210,0	-1518,0	114,60	NN	1	made ground (variable composition)
O-2/226	0,30	-1210,0	-1518,0	114,60	ML	12	silt
O-2/226	4,00	-1210,0	-1518,0	114,60	SW-SC	14	clayey sand
O-2/226	4,60	-1210,0	-1518,0	114,60	SM	16	silty sand
O-2/226	5,40	-1210,0	-1518,0	114,60	SP	16	fine grained sand with clay interbeddings
O-2/226	6,20	-1210,0	-1518,0	114,60	ML	16	silt
O-2/226	7,00	-1210,0	-1518,0	114,60	ML	16	silty sand

Structure of geological data used for computations is presented in the table 1. Each record in the table represents a roof depth of unique lithological layer documented in each borehole. The coordinates of each layer roof depth point in the borehole profile are spatialy localized by x,y,z coordinates, given in the local coordinate system (Warszawa 75). Each lithological layer has its symbol (an standardized abbreviation) and a short description. The numbers of engineering-geological layers are ascribed to each lithological layer. It can be seen as an example, that in the table 1., all lithological layers from the depth of 4,60 m b.g.l. down to de end of borehole, are classified as engineering-geological layer no.16.

3   Computational Intelligence Methods for Engineering-Geological Layer Modeling

Supervised learning is machine learning, where an algorithm is presented with a training set T being a collection of training examples (instances, patterns, data points). Each instance in the training set is described by a pair the (x_i, y_i) where x_i is a vector of features and y_i is a label or a target of x_i. During the learning phase, a supervised machine learning algorithm searches for mapping x to y. During the classification phase additional instances are presented and the algorithm, based on the mapping found in the learning phase, estimates what target will be assigned to those new samples.

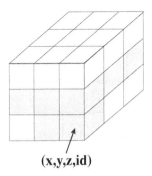

(x,y,z,id)

Fig. 4. Volumetric representation of space

The models are based on the volumetric representation of space. Each voxel is described by its coordinates and identifier of engineering-geological layer which it is belonging to (fig. 4).

Preparation of the layer model consists of following stages:

- Creation of learning data set - the available data (described in section 2) after removal of the artifacts is transformed (e.g. normalized) for use with the classifier.
- Learning of the classifier –neural network, in case of k-nn this stage is skipped.
- Classification of the voxels in considered area.
- A surface model can be prepared based on obtained volumetric data of considered layer.

Validation

In order to validate the classifiers false positive (FP), false negative (FN), true negative (TN) and true positive (TP), precision and recall [9, 10] parameters were calculated. Precision is defined as

$$precision = \frac{TP}{TP+FP} \cdot 100\% \qquad (1)$$

The recall is defined as

$$recall = \frac{TP}{TP+FN} \cdot 100\% \qquad (2)$$

Total accuracy is defined as

$$total\ accuracy = \frac{TP+TN}{TP+FP+TN+FN} \cdot 100\% \qquad (3)$$

Precision score of 100% for binary classification means that every item labeled as belonging to positive class does indeed belong to positive class. Recall of 100% means that every item from positive class was labeled as belonging to positive class.

Validation by the Experts
In order to enable validation of the results by the experts in the field of geology, cross-section were prepared of the obtained volumetric model. The shape of the terrain surface and the location of the section line is shown in fig. 5.

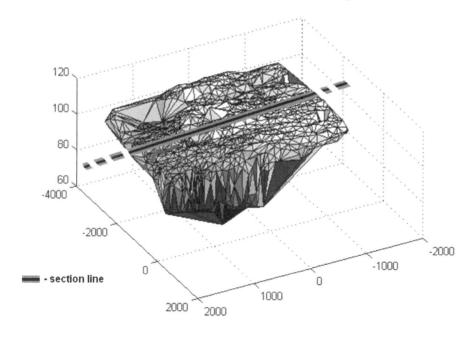

Fig. 5. The terrain surface obtained by Delaunay's triangulation of borehole data according to the engineering-geological profile. Vertices of triangles represent borehole locations. The model was base for cross section shown on figures. 6,7,8.

3.1   K-nn Model

The k-nearest neighbors (KNN) is an example of a supervised learning algorithm. It is one of the most widely used and well known classification algorithms. The KNN algorithm is used in wide range of classification problems such as: medical diagnosis, image processing, predicting of properties of amino acids sequences. The popularity of KNN is based on its simplicity and effectiveness.

The data set D is defined as:

$$D = \{(\mathbf{x}_i, t_i)\}, \quad \mathbf{x}_i \in X \subset R^d, \quad t_i \in \{-1, +1\} \tag{4}$$

An arbitrary data point $\mathbf{x}_i$ can be described by a vector $[x_{i1}, x_{i2}, \ldots, x_{id}]$, where d is the dimension of the samples space. The distance between two data points x_i and x_j is defined to be $dist(\mathbf{x}_i, \mathbf{x}_j)$, where

$$dist(\mathbf{x}_i, \mathbf{x}_j) \equiv \sqrt{\sum_{r=1}^{d} (x_{ir} - x_{jr})^2} \tag{5}$$

Let us first consider learning discrete-valued target functions of the form

$$f : R^n \rightarrow L \tag{6}$$

where L is the finite set of labels of data points $\{-1, 1\}$. During the training phase all labeled points are added to the list of training examples. The classification of data point $\mathbf{x}_q$ can be described as

$$f(x_q) \leftarrow \arg\max_{t \in L} \sum_{i=1}^{k} \delta(t, f(\mathbf{x}_i)) \tag{7}$$

where

$$\delta(a,b) = \begin{cases} 1 & a = b \\ 0 & a \ll b \end{cases}$$

and $\mathbf{x}_1, \ldots \mathbf{x}_k$ denote the k instances from training examples that are nearest to $\mathbf{x}_q$.

The k parameter (nearest neighbors count) was set to 5. Several experiments were made in order to determine the best value. It turned out that when the k value is increased the recall increases, but precision decreases. The original coordinates were scaled by factor $3.23 \cdot 10^4$ along X axis, $2.64 \cdot 10^4$ along Y axis and $1.49 \cdot 10^2$ along Z axis.

The results of KNN classification on the learning set and the test set are shown in table 2. The data set contained 5846 points. The learning set consisted of 5310 randomly selected points. 536 points were left out for testing purposes and were not included in the learning set. The cross-section of the obtained model is shown in fig. 6, which was used for validation by the geologists.

Table 2. Results of KNN classification of voxels (layer 15) on learning and test set

	Learning set	Test set
True positives (TP)	2616	167
True negatives (TN)	1936	196
False positives (FP)	704	82
False negatives (FN)	78	67
recall [%]	97,10	71,37
precision [%]	78,80	67,07
total accuracy [%]	85,00	70,90

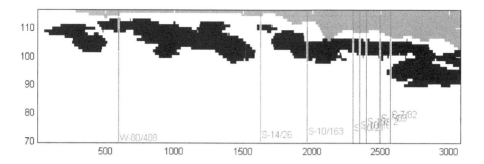

Fig. 6. Cross-section of obtained volumetric model (KNN classification). Starting point coordinate (-1200, -1000), end point coordinate (1880, -1000). Black – engineering-geological layer no. 15, white – other layers, grey – air. Vertical lines represent boreholes at a distance less than 10m from the cross-section line.

3.2   Neural Network Modeling

Neural network (multilayer perceptron) was used as engineering-geological layer classifier. The inputs receive 3-dimensional geometrical coordinates of a given point. Activation function of hidden neuron is sigmoidal (hyperbolic tangent) and output neuron is linear. The hidden layer consists of 200 neurons and output layer consists of 1 neuron. The classifier performs binary classification. The training examples (voxels) of a selected layer are labeled as +1, while the remaining are labeled as -1. The Levenberg-Marquardt algorithm was used for network training.

The separating surface of a given layer is a combination of planes created by the hidden neurons. Since an output neuron computes a weighted sum of its inputs, it can be seen as combining the planes represented by the hidden layer neurons [1]. The weights of a hidden neuron represent normal vector to the plane created by it. Hence, the neural model complexity equals to the number of hidden neurons, it increases as the area of approximation extends. The neural model covers whole input data domain, so this is a global model.

The results of neural network classification on the learning set and the test set are shown in table 3. The data set contained 5846 points. The learning set consisted of 5310 randomly selected points. 536 points were left out for testing purposes and were not included in the learning set. The accuracy of classification enables creation of volumetric models. Selected cross-sections of the models are shown in fig. 7,8, which were used for validation by geologists.

Table 3. Results of neural network classification of voxels (layer 15) on learning and test set

	Learning set	**Test set**
True positives (TP)	2391	186
True negatives (TN)	2009	214
False positives (FP)	625	70
False negatives (FN)	285	66
recall [%]	89,35	73,81
precision [%]	79,28	72,66
total accuracy [%]	82,86	74,63

The results obtained by the KNN and neural classifier on the test set are comparable. The KNN performs much better on the learning set due to local nature of this model. The shapes of the engineering-geological layers obtained by the neural classifiers are smoother than shapes generated by the KNN method (fig. 6,8). A significant difference between the models lies is the time needed for creation of the model and classification of an unknown sample. The creation of a KNN model requires only collecting of the data set containing example points and their labels. The neural classifier needs an additional step (computationally intensive) consisting of calculation of the neurons weights. However, once the neural model is ready, the prediction is very fast. The KNN classifier requires a data search for each predicted point.

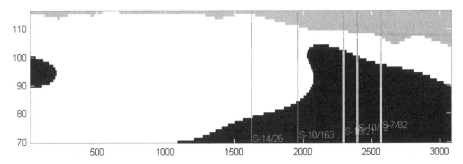

Fig. 7. Cross-section of obtained volumetric model (neural network classification). Starting point coordinate (-1200, -1000), end point coordinate (1880, -1000). Black – engineering-geological layer no. 27, white – other layers, grey – air. Vertical lines represent boreholes at a distance less than 10m from the cross-section line.

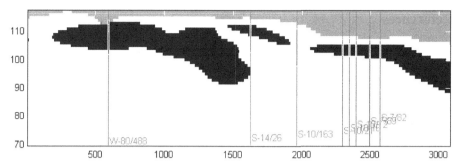

Fig. 8. Cross-section of obtained volumetric model (neural network classification). Starting point coordinate (-1200, -1000), end point coordinate (1880, -1000). Black – engineering-geological layer no. 15, white – other layers, grey – air. Vertical lines represent boreholes at a distance less than 10m from the cross-section line.

4   Conclusions

The use of volumetric methods for modeling of engineering-geological layers is not a common approach. According to the results obtained from analyses performed in this paper, the voxel classification method offers attractive possibilities to model complex, discontinuous shapes, characteristic to geological layers. This method enables predicting of engineering-geological layers in unmeasured areas between boreholes. The results can be easily visualized and presented to the experts for validation.

Method of neural network classification was applied to model the surfaces of engineering-geological layers no. 15 (pleistocene fluvioglacial sands) as a discontinuous layer and 27 (pliocene clays) as a continuous discrete layer. The shapes generated with the use of the neural classification are gently and irregularly undulated. This is a good form of a representation especially for glacitectonicaly deformed surfaces of pliocene clays from area of Warsaw. Also the effect of modeling of the discontinuous layer of fluvioglacial sands is interesting. The shapes of sand lenses generated by the model seam to represent the geological conditions in a appropriate way.

The effects of volumetric modeling with supervised learning methods can be presented in form of cloud of points generated by the presented in the paper numerical algorithms. Such data can be then easily processed in different visualization environments, like GIS software platforms or geological 3D modeling software. The geological model based on data processed by methods like neural approximation and k-nn can be used to generate cross-sections and engineering-geological maps at certain, user defined depths.

Concluding, the presented methods can be assessed as a useful tool for engineering-geological layers modeling, with broad scope of practical applications.

References

1. Dreyfus, G.: Neural Networks - Methodology and Applications. Springer, Heidelberg (2005)
2. Cover, T.M., Hart, P.E.: Nearest Neighbor Pattern Classification. IEEE Transactions on Information Theory 13(1), 21–27 (1967)
3. Kulkarni, S.R., Lugosi, G., Venkatesh, S.S.: Learning Pattern Classification—A Survey. IEEE Transactions on Information Theory 44(6), 2178–2206 (1998)
4. Shakhnarovich, G., Darrell, T., Indyk, P. (eds.): Nearest-Neighbor Methods in Learning and Vision. MIT Press (2006)
5. Alippi, C., Fuhrman, N., Roveri, M.: k-NN classifiers: investigating the k=k(n) relationship. In: Proc. 2008 International Joint Conference on Neural Networks (IJCNN 2008), pp. 3676–3680 (2008)
6. Hastie, T., Tibshirani, R., Friedman, J.: The Elements of Statistical Learning - Data Mining, Inference, and Prediction. Springer (2008)
7. Mitchell, T.M.: Machine Learning. McGraw-Hill Science/Engineering/Math (1997)
8. Kraiński, A., Mrówczyńska, M.: The use of neural networks in the construction of geological model of the Głogów-Baruth Ice-Marginal Valley in the Nowa Sól area, Poland. Prz. Geol. 60, 650–656 (2012)
9. Makhoul, J., Kubala, F., Schwartz, R., et al.: Performance measures for information extraction. In: Proc. DARPA Broadcast News Workshop, Herndon, VA (1999)
10. van Rijsbergen, C.V.: Information Retrieval. Butterworth, London (1975)
11. Kumar, J., Konno, M., Yasuda, N.: Subsurface Soil-Geology Interpolation Using Fuzzy Neural Network. J. Geotech. Geoenviron. Eng. 126(7), 632–639
12. Mohseni-Astani, R., Haghparast, P., Bidgoli-Kashani, S.: Assessing and Predicting the Soil Layers Thickness and Type Using Artificial Neural Networks - Case Study in Sari City of Iran. Middle-East Journal of Scientific Research 6(1), 62–68 (2010)

A Hybrid Genetic Algorithm and Radial Basis Function NEAT

Heman Mohabeer and K.M. Sunjiv Soyjaudah

Department of Electrical and Electronics Engineering
University of Mauritius
Le Reduit, Moka, Mauritius
heman.mohabeer@ieee.org, ssoyjaudah@uom.ac.mu

Abstract. We propose a new neuroevolution technique that makes use of genetic algorithm to improve the task provided to a Radial Basis Function – NEAT algorithm. Normally, Radial Basis Function works best when the input-output mapping is smooth, that is, the dimensionality is high. However, if the input changes abruptly, for example, for fractured problems, efficient mapping cannot happen. Thus, the algorithm cannot solve such problems effectively. We make use of genetic algorithm to emulate the smoothing parameter in the Radial Basis function. In the proposed algorithm, the input- output mapping is done in a more efficient manner due to the ability of genetic algorithm to approximate almost any function. The technique has been successfully applied in the non-Markovian double pole balancing without velocity and the car racing strategy. It is shown that the proposed technique significantly outperforms classical neuroevolution techniques in both of the above benchmark problems.

Keywords: genetic algorithm, radial basis function, car racing strategy, double pole balancing.

1 Introduction

Learning is regarded as an integral form of contribution toward better evolution. An agent learns to react faster and more efficiently to change its external environment during its lifetime. Whereas evolution produces phylogenetic adaptation, lifetime learning guides evolution to higher fitness [1]. This lifetime learning process is known as the Baldwin effect [2] [3]. In order to design much better adaptive artificial intelligence (AI) systems, a study of the above two concepts i.e. evolution and learning, needs to be performed.

Evolution alone cannot cater for changes in environment since it is relatively a slow process. An agent needs to be always active to any change that occurs in its environment in order to maintain its performance. This is achieved by constant learning.

In recent years, the interest in machine learning has significantly increased with the advent of new algorithms capable of evolving their behaviors based on empirical data. Neuroevolution (NE) is one of the forms of machine learning that uses evolutionary algorithms [4] [5] [6] to evolve artificial neural network (ANN).

V. Golovko and A. Imada (Eds.): ICNNAI 2014, CCIS 440, pp. 159–170, 2014.

One way in which agents based on ANN can evolve their behavior is allowing them to change their integral structure, thereby exhibiting plasticity. In this way, they follow the same concept of real organisms toward changing and unpredictable environment [7] [8] [9]. In 2002, Stanley et al. developed neuroevolution of augmented topologies (NEAT) [10], which makes use of genetic algorithms (GA) to evolve. Although not new [11] [12], the idea of speciation to allow historical marking was introduced in a more efficient manner [10] in order to optimize the functionality of NEAT. Benchmark problems such as the pole balancing and the double pole balancing have been solved with greater efficiency and lower complexity as compared to conventional NE and evolutionary programming [13] [14].

Even though NEAT has performed well on many reinforcement learning problems [15] [16] [18] yet, on a high level domain control such as the racing strategy [19], NEAT has performed rather poorly. One approach to solve these kinds of problems was to adopt the concept of the fractured domain hypothesis. Kohl et al. [30] successfully improved the performance of NEAT by using radial basis function (RBF) instead of the classical multilayer perceptron (MLP). It was reported in [20] that a value- function reinforcement learning method normally benefits from approximating value function using a local function approximator like RBF networks. Stone et al. used the benchmark keep away soccer domain to demonstrate that a RBF- based value function approximator allowed local behavioral adjustments [22] [23] thus outperforming normal neural network value function approximator.

Such results were promising for neuroevolution learning. Although, this method was successful, it did not provide a plausible explanation as to why NEAT performed poorly on fractured domains. This paper introduces a new method which makes use of GA to help RBF-NEAT solve reinforcement learning problems. To demonstrate the potential of this approach, this paper performs comparisons in the racing strategy domain [19] and the double pole balancing problem.

By allowing GA to emulate a smoothing operator in a given problem prior to its exposure to the RBF-NEAT algorithm, the performance of the algorithm significantly increases. In fact, it outperforms the traditional RBF-NEAT in fractured domain, suggesting a powerful new means of solving high level behavior as well as conventional reinforcement learning tasks [15].

The remainder of the paper is organized as follows: Section 2 provides a brief methodology of the theories involved. Section 3 consists of the methodology adopted to implement the proposed system. Section 4 discusses the results obtained and comparisons are made with various other algorithms used to solve the same problem. Finally, section 5 concludes the proposed work and discusses future works that may arise as a result of the proposed work.

2   Reinforcement Learning

Reinforcement learning (RL) is concerned with how an agent (actor) ought to take action in an environment so as to maximize some notion of cumulative reward. RL learn policies for agents acting in an unknown stochastic world, observing the state that occur and the rewards that are given at each step [24]. In this way the agent progresses towards the desired solutions iteratively. Reinforcement learning problems are divided into two categories, namely low level reinforcement learning and high level reinforcement learning problem. One example of a low level RL problem is the non- Markovian double pole balancing which is discussed below, while the racing strategy is presented as a high level RL problem. Both of these problems are used later as benchmarks to test the proposed algorithm.

2.1 Non- Markovian Double Pole Balancing (DPNV)

Non- Markovian double pole balancing (DPNV) [25] can be considered as a difficult benchmark task for control optimization. K. Stanley et al. [10] compared the results of the neuroevolution methods which have reportedly solved the DPNV problem: Cellular Encoding CE [26], Enforced Sub Populations ESP [27], and NEAT outperformed all the other algorithms.

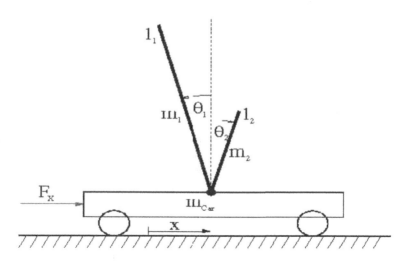

Fig. 1. Double Pole Balancing

The double pole balancing setup (see figure 1), consists of a car with mass (m_{car}) 1 kg and one degree of freedom x, in which two poles of different lengths $l1 = 1$ m and $l2 = 0.1$ m are mounted. The poles have the masses $m1 = 1$ kg and $m2 = 0.1$ kg. Based on the measured values of the joint angles $\theta1 = \theta2$ and the position of the car x, the controller is required to balance both of the poles by applying a force F_x (with a maximal magnitude $F_{max} = 10$ [N]). Assuming rigid body dynamics and neglecting friction, the system can be described by the equations of motion shown below.

Note: $\ddot{x}$ = acceleration of the cart (second derivative of x with respect to time)

N = number of poles (in this paper, it is assigned a value of 2)
Fi = effective force from the ith pole on the cart,
θ_i = angle of the ith pole from the vertical
l_i = half length of the pole

$$= \frac{F_{x-\sum_{i=1}^{N} \widetilde{F_l}}}{M_{car} + \sum_{i=1}^{N} \widetilde{m_l}}$$

$$\ddot{\theta} = -\frac{3}{4l}\left(x \ddot{\cos} \theta_i + g \sin \theta_i\right)$$

The numerical simulation of the system is based on a 4th-order Runge-Kutta integration of these equations with a time step of$\Delta t = 0.01s$. The above parameters and setups were inspired from [29] as the same configurations were adopted for comparison with other neuroevolution method cited above.

2.2   Racing Strategy Domain

The racing strategy (RS) domain is considered as a high level reinforcement learning problem. A more ample description has been given in [20]. RS domain has a continuous state space, meaning that it is much like real world RL tasks. First introduced to the artificial intelligence community by [31] [32], it has since then garnered lots of interest. The goal of the racing strategy domain implemented in this paper to a network that makes this decision on which waypoint the agent should focus. The aim to ensure that each waypoint may only be awarded to one of the two players, and after a fixed amount of time the player that has "collected" the most waypoints wins. The implementation of the racing car strategy was inspired from that implemented in [34]. The source code was obtained from [33]. The following requirements were observed to ensure that the car racing domain is effectively implemented as a high level reinforcement learning task.

- The capability of generating a large number of random tracks with very little efforts.
- The model must be fast to compute. Evolutionary algorithms may require millions of simulated time steps in order to converge.
- The sensor inputs for a controller should be reasonably simple; this encourages more members of the research community to participate.
- The waypoints are randomly distributed around a square area at the beginning of each race, and the car knows the position of the current waypoint and the next waypoint.

The simulations result for both the double pole balancing and the car racing strategy are discussed later in this paper. As part of the buildup of the proposed algorithm, the next section gives a vague description of GA and its contributions in neuroevolution.

2.3   Genetic Algorithms

Genetic algorithm (GA) has been inspired by the Darwinian theory of evolution. Basically, GA consists of any population based model that uses selection and recombination operators to generate new sample points in a search space. Crossover and mutation are the two major reproductive operators that are responsible to evolving GA toward much fitter generation.

Crossover, in the context of GA, is the process of combining two chromosomes to produce new offspring. The idea is to transmit the best characteristics from each parent chromosomes to the offspring. In this way, the new generation of individuals will be more efficient than the previous one. A crossover operator dictates the way selection and recombination of the chromosomes occur. Moreover, these operators are usually applied probabilistically according to a crossover rate.

Mutation refers to the random process changing an allele of a gene in order to produce a new genetic structure. The probability of mutation is usually applied in the range of 0.001 to 0.01, and modifies elements of the chromosomes. It has been reported in [34] that mutation is responsible for preventing all the solutions in a population from falling into a local optimum of solved problem. It is important to note that mutation is both dependent on crossover and the encoding scheme.

To summarize, it could be said that the role of mutation consists of restoring lost or unexplored genetic material into the populations. A fitness value is assigned to each chromosome such that the one which is fit is allowed to crossover. The fitness function is responsible for assigning each chromosome a fitness value based on their performance in the problem domain. It is imperative to design a good fitness function in order to effectively solve a problem. A good fitness function will help in probing the search space more efficiently and effectively. It also helps to escape local optimum solution. Across each generation of population there is a convergence toward an overall higher fitness value.

In this way, GA probes for the solution in the search space. Speciation is another property which has been given significant amount of importance in NEAT. Speciation has the main function of preserving diversity. In NEAT, it is used to protect innovation thus speciation allows organisms to compete primarily within their own niches instead of with the population at large. Topological innovations are protected in a new niche where they have time to optimize their structure through competition within the niche [17]. One of the most attracting features of GA is its robustness and efficacy. A more in depth analysis of GA is provided in [12].

3   Methodology

Figure 2 displays a schematic approach of the proposed algorithm. GA is deployed at two instances in the diagram. At the first instance, GA is deployed as the first step during which the algorithm is deployed to a benchmark problem. For example, consider the case for the racing car strategy. An attempt is made to solve the problem thus the random population generated by the GA is directed to the problem. A fitness value is assigned as a stopping criterion such that when this value is reached, GA instantly stops probing that state space. It is assumed that during the search, GA allows the problem to be either solved or develop a pattern which is directed to the solution is created. Both possibilities should be considered since depending on the complexity of the problem it is postulated that GA may or may not find the optimal solution on its own.

Hence after an initial pass through GA, the pattern developed within the population enables NEAT to converge toward the solution in a more efficient manner. The algorithm has been implemented using matlab software. The initial NEAT software was obtained from [33].

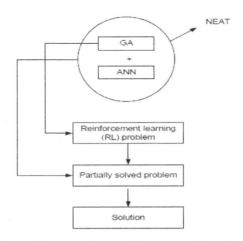

Fig. 2. Schematic diagram of the proposed algorithm

Figure 2 provides an overview of the approach taken to implement the algorithm. The population size for both NN and the reinforcement learning tasks was set to a fixed value of 200. This value was taken as default from the original NEAT algorithm. During the implementation phase, the most important aspect was to ensure that the GA was reinitialized when switching from the RL task to the NN. This primary reason for this precaution was to ensure that the population sample from the network and the RL task are not mix since they are both encoded in the same way.

The crossover rate and the mutation rate have been fixed at 0.8 and 0.01 respectively as per the theoretical value implemented in the NEAT algorithm. The RBF nodes are initially set to the lowest level. Whenever a new node is added, the weights of the existing nodes in the network are frozen to focus the search process on the most recently added node. In this way, the frozen weights retain some information, which at a later stage could be retrieving successfully. GA switches from one population, that is, from the RL task to the NN in a very simplistic manner. A fitness value is assigned to each population, which acts as a stopping criterion and thus activates the switch from one population to the other.

In the proposed algorithm, the maximum overall fitness value of 16 has been assigned to both set of population. As stated previously, this value acts as a stopping criterion for the algorithm to complete searching within the search space. All the parameters set has been standardized so that the algorithm could be compared and contrasted to known reported results.

4   Results and Discussion

The RBF network for the DPNV problem was implemented with six input units and one output unit. Table 1 displays results after simulations of the RBF network, NEAT, and the proposed algorithms. The evaluations refer the number of iterations performed prior to reaching the best performance while the neural network refers to the complexity of the patterns and the amount of NN needed to achieve the optimal performance.

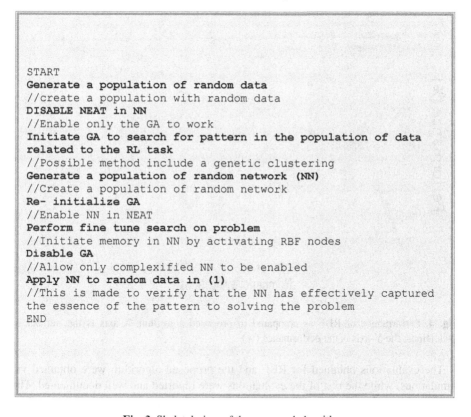

```
START
Generate a population of random data
//create a population with random data
DISABLE NEAT in NN
//Enable only the GA to work
Initiate GA to search for pattern in the population of data
related to the RL task
//Possible method include a genetic clustering
Generate a population of random network (NN)
//Create a population of random network
Re- initialize GA
//Enable NN in NEAT
Perform fine tune search on problem
//Initiate memory in NN by activating RBF nodes
Disable GA
//Allow only complexified NN to be enabled
Apply NN to random data in (1)
//This is made to verify that the NN has effectively captured
the essence of the pattern to solving the problem
END
```

Fig. 3. Skeletal view of the proposed algorithm

Table 1. Comparisons of Various Neuroevolution Techniques

Algorithm	Evaluations	Neural Nets
CE	840,00	1600
ESP	169, 46	1000
NEAT	33,184	1000
RBF	33,275	
Proposed algorithm	14,237	350

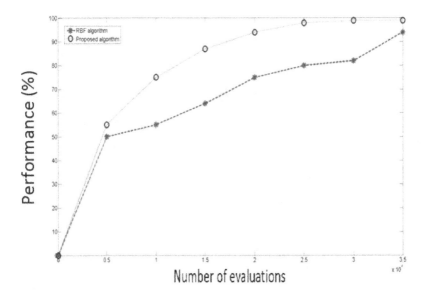

Fig. 4. Performance of RBF as compared to proposed algorithm, X-axis is the number of evaluations; the Y-axis is the performance (%)

The evaluations obtained for RBF and the proposed algorithm were obtained via simulations, while the rest of the evaluations were reported and well documented. The results obtained were averaged upon 100 simulations. Figure 4 compares the simulation results for RBF and the proposed algorithm. The performance of the proposed algorithm is significantly better. The performance has been generated upon the successful solving of the DPNV problem with a higher level of accuracy. However, a performance of greater than 80% has been considered to be good enough to be regarded as successful. Both curves roughly undergo the same progress.

Conversely, it is postulated that GA in the proposed algorithm cater for the improved performance. The proposed algorithm reports a smaller average number of function evaluations to solve the problem. Driven by GA, it is able to solve the problem much easily since GA tends to reduce the state space by partially re-adapting the problem for a much easier evaluation of the subsequent algorithm. The second simulations were for a high level domain control problem. In this paper, the car racing strategy has been proposed. The experiment consisted of 15 runs and each run consisted of a race of 1400 laps. Fifteen runs were performed as the after this amount not much variation were detected in the results.

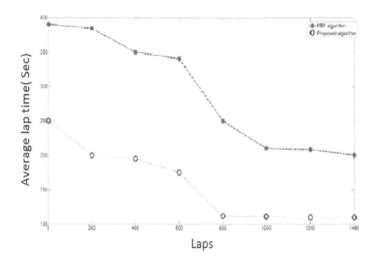

Fig. 5. Time taken to complete each lap for the 1400 laps

Figure 5 compared the time taken to complete a lap for the 1400 laps. ESP and CE had no evaluation record for the car racing strategy domain. Therefore, comparisons were done between RBF and the proposed algorithm. It was observed that the proposed algorithm were immediately more successful than RBF. One plausible explanation is the fact that, like in the DPNV problem, GA contributes a lot in familiarizing the data to the subsequent algorithm. In this way, it gives a significant advantage as compared to other traditional neuroevolution techniques such as the NEAT and even the RBF network.

The proposed algorithm has been constructed using two powerful, well known algorithms; GA and RBF. Unlike NEAT, it makes use of GA not only to evolve the RBF network, but also in an attempt to reorganize the problem such that upon complexification, the RBF network is exposed to a date which is partially solved. In this way, it has a definite advantage over other neuroevolution techniques since it will require a lesser complex network and secondly, it will tend to solve a problem in a faster and more efficient way. Higher level domain problems such as the car racing strategy are considered as a difficult problem that requires much complexity. In this paper, this complexity is reduced by GA. NEAT has to undergo a full complexification in order to solve such problem while the proposed algorithm is exposed to "treated" data. High level domain problems are considered as fractal in nature [33] local processing in RBF seems to be able to do much better in these situations.

It has been reported in [36] that RBF performs best when approximating a smooth input output mapping. In this paper, the most logical explanation for the significant improvement brought by the proposed algorithm lies on the fact that the GA enables smoothening of input data to the RBF-NEAT. In this way, the output rendered is optimal since a lesser complexity is achieved by NEAT. This results in the problems depicted to be solved much quicker.

5   Conclusion and Future Work

In this work, we introduce a technique that makes use of the capabilities of GA to achieve an optimal state using lower complexity in RBF- NEAT when attempting to solve benchmark problems. The problems to which the proposed algorithm was exposed represented both the low level and high level domains. Moreover, comparison has been drawn with classical neuroevolution techniques. It was shown that the new algorithm outsmarted those techniques quite significantly.

The complexity of NN in NEAT was lower as compared to conventional algorithms (as shown in the table above). This is important as it reduces time for solving the problems and also makes use of lesser resources in terms of computation. GA is known to be a robust algorithm. We have extended its capabilities by allowing it to complexify the RBF-network and smoothen the input for the network to work in a more effective manner. It is worthwhile to note that the algorithm has been successfully applied to reinforcement learning problems. An interesting aspect would be to note its behavior in real life problems such consisting of fractal domains.

Acknowledgements. The financial support of the Tertiary Education of Mauritius is gratefully acknowledged.

References

1. Karpov, I., Sheblak, J., Miikkulainen, R.: OpenNERo; A game platform for AI research and education. In: Proceeding of the Fourth Artificial Intelligence and Interactive Digital Entertainment Conference (2008)
2. Shettleworth, S.J.: Evolution and Learning-The Baldwin effect reconsidered. MIT Press, Cambridge (2003)
3. Kull, K.: Adaptive Evolution Without Natural Selection-Baldwin effect. Cybernetics and Human Knowing 7(1), 45–55 (2000)
4. Baeck, T.: Evolutionary Algorithms in Theory and Practice. Oxford University Press, New-York (1996)
5. Beyer, H.G.: The Theory of Evolution Strategies. Springer, Heidelberg (2001)

6. Yao, X., Liu, Y.: A new Evolutionary System for Evolving Artificial Neural Networks 8(3), 694–713 (1997)
7. Floreano, D., Urzelai, J.: Evolutionary Robots with Self-Organization and Behavioral Fitness. Neural Networks 13, 431–443 (2000)
8. Niv, Y., Noel, D., Ruppin, E.: Evolution of reinforcement learning in Uncertain Environments; A Simple Explanation for Complex Foraging Behaviors. Adaptive Behavior 10(1), 5–24 (2002)
9. Soltoggio, A., Bulliaria, J.A., Mattiussi, C., Durr, P., Floreano, D.: Evolutionary Advantages of Neuromodulated Plasticity in Dynamics, Reward based Scenario. In: Artificial Life XI, pp. 569–576. MIT Press, Cambridge (2008)
10. Stanley, K.O., Miikkulainen, R.: Evolving Neural Networks through Augmenting Topologies. Evolutionary Computation 10(2) (2002)
11. Potter, M.A., De Jong, K.A.: Evolving Neural Networks with Collaborative Species. In: Proceedings of the 1995 Summer Computing Simulation Conference (1995)
12. Radcliff, N.J.: Genetic Set Recombination and its application to Neural Network Topology Optimization. Neural Computing and Applications 1(1), 67–90 (1992)
13. Wieland, A.: Evolving Neural Networks Controllers for unstable systems. In: Proceedings of the International Joint Conference on Neural Networks, Seattle, WA. IEEE (1991)
14. Saravanan, N., Fogel, D.B.: Evolving Neural Control Systems. IEEE Expert, 23–27 (1995)
15. Reisinger, J., Bahceci, E., Karpov, I., Miikkulainen, R.: Coevolving Strategies for general game playing. In: Proceedings of the IEEE Symposium on Computational Intelligence and Games (2007)
16. Stanley, K.O., Bryant, B.D., Miikkulainen, R.: Real-Time Neuroevolution in the NERO video game. IEEE Transactions on Evolutionary Computation 9(6), 653–668 (2007)
17. Stanley, K.O., Miikkulainen, R.: Competitive Coevolution through Evolutionary Complexification. Journal of Artificial Intelligence Research, 63–100 (2004)
18. Stanley, K.O., Miikkulainen, R.: Evolving a roving eye for go. In: Proceeding of the Genetic and Evolutionary Computation Conference (2004)
19. Lucas, S.M., Togelius, J.: Point-to-pointcar racing; an Initial Study of Evolution Versus Temporal Difference Learning. In: IEEE Symposium of Computational Intelligence and Games, pp. 260–267 (2007)
20. Li, J., Martinez-Maron, T., Lilienthal, A., Duckett, T.: Q-ran; A Constructive Reinforcement Learning approach for Robot behavior Learning. In: Proceedings of IEEE/RSJ International Conference on Intelligent on Intelligent Robot and System (2006)
21. Stone, P., Kuhlmann, G., Taylor, M.E., Liu, Y.: Keepaway Soccer: From Machine Learning Testbed to Benchmark. In: Bredenfeld, A., Jacoff, A., Noda, I., Takahashi, Y. (eds.) RoboCup 2005. LNCS (LNAI), vol. 4020, pp. 93–105. Springer, Heidelberg (2006)
22. Moody, J., Darken, C.J.: Fast Learning in Networks of Locally tuned Processing units. Neural Computation, 281–294 (1989)
23. Sutton, R.S., Barto, A.G.: Reinforcement Learning; An Introduction. MIT Press (1998)
24. Wieland, A.: Evolving Neural Network Controllers for Unstable Systems. In: Proceedings of the IJCNN, Seattle, WA, pp. 667–673. IEEE (1991)
25. Gruau, F., Whitley, D., Pyeatt, L.: A comparison between Cellular Encoding and Direct Encoding for Genetic Programming. In: Genetic Programming 1996: Proceedings of the First Annual Conference, pp. 81–89 (1996)
26. Gomez, F.J., Miikkulainen, R.: Solving Non-Markovian Control tasks with Neuroevolution. In: Proceedings of the IJCAI, pp. 1356–1361 (1999)

27. Dürr, P., Mattiussi, C., Floreano, D.: Neuroevolution with Analog Genetic Encoding. In: Runarsson, T.P., Beyer, H.-G., Burke, E.K., Merelo-Guervós, J.J., Whitley, L.D., Yao, X. (eds.) PPSN IX. LNCS, vol. 4193, pp. 671–680. Springer, Heidelberg (2006)

28. Jakobsen, M.: Learning to Race in a Simulated Environment, http://www.hiof. no.neted/upload/attachment/site/group12/Morgan_Jakobsen_Lear ning_to_race_in_a_simulated_environment.pdf (last accessed March 18, 2012)

29. Kietzmann, T.C., Reidmiller, M.: The Neuro Slot car Racer; Reinforcement Learning in a Real World Setting, http://ml.informatik.unifreiburg.de/_media/ publications/kr09.pdf (last accessed March 12, 2013)

30. Kohl, N., Miikkulainen, R.: Evolving Neural Networks for Fractured Domains. Neural Networks 22(3), 326–337 (2009)

31. Togelius, J., Lucas, S.M.: IEEE CEC Car Racing Competition, http://Julian. togelius.com/cec2007competition/ (last accessed February10, 2012)

32. NEAT Matlab, http://www.cs.utexas.edu/users/ai-lab/?neatmatlab (last accessed March 02, 2012)

33. The Radial Basis Function Network, http://www.csc.kth.se/utildning/ kth/kurser/DD2432/ann12/forelasningsanteckningar/RBF.pdf (last accessed: June 21, 2012)

34. Bajpai, P., Kumar, M.: Genetic Algorithm-an Approach to Solve Global Optimization Problems. Indian Journal of Computer Science and Engineering 1(3), 199–206 (2010)

A Multi-agent Efficient Control System for a Production Mobile Robot

Uladzimir Dziomin[1], Anton Kabysh[1], Vladimir Golovko[1], and Ralf Stetter[2]

[1] Brest State Technical University
[2] University of Ravensburg-Weingarten

Abstract. This paper presents the results of the experiments of a multi-agent control architecture for the efficient control of a multi-wheeled mobile platform. Multi-agent system incorporates multiple Q-learning agents, which permits them to effectively control every wheel relative to other wheels. The learning process was divided into two steps: *module positioning* – where the agents learn to minimize the error of orientation and *cooperative movement* – where the agents learn to adjust the desired velocity in order to conform to the desired position in formation. From this decomposition every module agent will have two control policies for forward and angular velocity, respectively. The experiments were carried out with a real robot. Our results indicate the successful application of the proposed control architecture for the real production robot.

Keywords: control architecture, multi-agent system, reinforcement learning, Q-Learning, efficient robot control.

1 Introduction

An efficient robot control is an important task for the applications of a mobile robot in production. The important control tasks are power consumption optimization and optimal trajectory planning. Control subsystems should provide energy consumption optimization in a robot control system. Four levels of robot power consumption optimization can be distinguished:

1. *Motor power consumption optimization.* Those approaches based on energy-efficient technologies of motor development that produce substantial electricity saving and improve the life of the motor drive components [1], [2].
2. *Efficient robot motion.* Commonly, this is a task of an inverse kinematics calculation. But the dynamic model is usually far more complex than the kinematic model [3]. Therefore, intellectual algorithms are relevant for the optimization of a robot motion [4].
3. *Efficient path planning.* Such algorithms build a trajectory and divide it into different parts, which are reproduced by circles and straight lines. The robot control subsystem should provide movement along the trajectory parts. For example, Y. Mei and others show how to create an efficient trajectory using

V. Golovko and A. Imada (Eds.): ICNNAI 2014, CCIS 440, pp. 171–181, 2014.

knowledge of the energy consumption of robot motions [5]. S. Ogunniyi and M. S. Tsoeu continue this work using reinforcement learning for path search [6].

4. *Efficient robot exploration.* When a robot performs path planning between its current position and its next target in an uncertain environment, the goal is to reduce repeated coverage [7].

The transportation of cargo is an actual task in modern production. Multi-wheeled mobile platforms are increasingly being used in autonomous transportation of heavy components. One of these platforms is a production mobile robot, which was developed and assembled at the University of Ravensburg-Weingarten, Germany [3]. The robot is illustrated in Figure 7.1a. The platform dimensions are 1200cm in length and 800cm in width. The maximum manufacturer's payload is 500kg, battery capacity is 52Ah, and all modules drive independently.

The platform is based on four vehicle steering modules [3]. The steering module (Fig. 7.1b) consists of two wheels powered by separate motors and behaves like a differential drive.

The goal is to achieve a circular motion of a mobile platform around a virtual reference beacon with optimal forward and angular speeds. One solution to this problem [8]-[10] is to calculate the kinematics of a one-wheeled robot for circle driving and then generalize it for multi-vehicle systems. This approach has shown promising modeling results. The disadvantage of this technique is its low flexibility and high computational complexity.

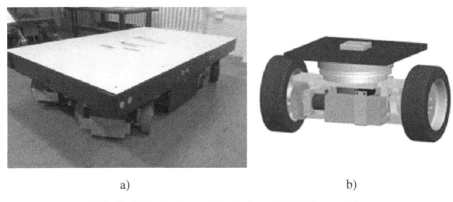

a)   b)

Fig. 1. a) Production mobile platform; b) Driving module

An alternative approach is to use the machine learning theory to obtain an optimal control policy. The problem of multi-agent control in robotics is usually considered as a problem of formation control, trajectory planning, distributed control and others. In this paper we use techniques from multi-agent systems theory and reinforcement learning to create the desired control policy.

The key contribution of this paper is an application of the multi-agent reinforcement learning approach for the efficient control of an industrial robot. The advantages of the approach were proven by the experiments in a laboratory with the real robot. They will be shown in Chapter 6 of the paper.

2   Steering Module Agent

Let's decompose the robot's platform into the independent driving module agents. The agent stays in physical, 2-D environment with a reference *beacon*, as shown in Fig. 2. The beacon position is defined by coordinates (x_b, y_b). The rotation radius ρ is the distance from the center of the module to the beacon.

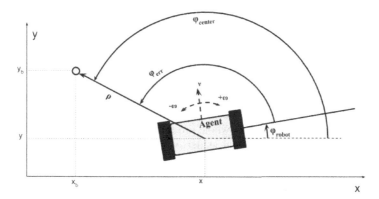

Fig. 2. State of the module with respect to reference beacon

In the simulated model environment, all necessary information about an agent and a beacon is provided. In a real robotic environment, this information is taken from wheel odometers and a module angle sensor. The environment information states are illustrated in Table 1. The navigation subsystem of real steering uses odometer sensors for navigation purposes in the presented platform. The full set of actions available to the agent is presented in Table 2. The agent can change the angle error φ_{err} around beacon, using control of linear v and angular speed ω.

Table 1. Environmental Information

№	Robot Get	Value
1	X robot position, x	Coordinate, m
2	Y robot position, y	Coordinate, m
3	X of beacon center, x_b	Coordinate, m
4	Y of beacon center, y_b	Coordinate, m
5	Robot orientation angle, φ_{robot}	Float number, radians $-\pi < \varphi_{robot} \leq \pi$
6	Beacon orientation angle relative to robot, φ_{center}	Float number, radians $-\pi < \varphi_{center} \leq \pi$
7	Radius size, ρ	Float number, m

Table 2. Agent actions

№	Robot actions	Value
1	Increase force, v_+	+0.01 m/s
2	Reduce force, v_-	-0.01 m/s
3	Increase turning left, ω_+	+0.01 rad/s
4	Increase turning right, ω_-	-0.01 rad/s
5	Do nothing, $\varnothing$	0 m/s, 0 rad/s

3   Multi-agent System of Driving Modules

One solution of formation control is the virtual structure approach [11]. The basic idea is to specify a virtual leader or a virtual coordinate frame located at the virtual center of the formation as a reference for the whole group such that each module's desired states can be defined relative to the virtual leader or the virtual coordinate frame. Once the desired dynamics of the virtual structure are defined, then the desired motion for each agent is derived. As a result, single module path planning and trajectory generation techniques can be employed for the virtual leader or the virtual coordinate frame while trajectory tracking strategies can be employed for each module.

Let, N steering module agent's with virtual leader forms a multi-agent system called platform. Fig. 3 shows an illustrative example of such a structure with a formation composed of four modules, where (x_b, y_b) represents the beacon and C represents a virtual coordinate frame located at a virtual center (x_c, y_c) with an orientation φ_c relative to beacon and rotation radius ρ_c.

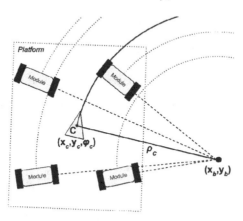

Fig. 3. The steering modules platform

Platform contains additional information such as square of platform and required module topology including its desired positions relative to the centroid of platform. Virtual leader seen by environment analogously to single steering module agent – it has a state, and can perform action. It receives the same information from environment defined in Table 1, and action set defined in Table 2. It should be noted, that modules not directly controlled by virtual leader. The modules remain independent entities and adopt their behavior to conform desired position in platform. In fig. 4, (x_i, y_i) and (x_i^{opt}, y_i^{opt}) represent, respectively, the i-th module's actual and desired position, and represent the desired deviation vector of the i-th module relative to desired position, where

$$\overline{d}_i^{err} = \overline{d}_i^t - \overline{d}_i^{opt} \tag{1}$$

Here $\overline{d}_i^t$ – distance from virtual center to current module position and $\overline{d}_i^{opt}$ required distance between virtual center and i-th module position derived from platform topology.

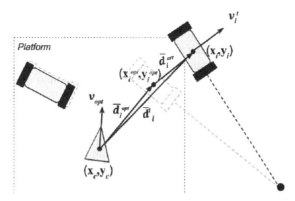

Fig. 4. State of the platform with reference to the i-th module

4   Module Positioning

The section discusses a reinforcement learning method producing efficient control law for module orientation around beacon.

The problem of finding multi-agent control law for circular motion can be decomposed in the following steps (a) *module positioning* and (b) *forward speed adjustment* to fit desired radius and position. For both problems we use reinforcement learning, which permits to achieve generalization ability. For example, the new beacon position can be dynamically assigned to platform, and the same control low can be used for module positioning.

4.1   Reinforcement Learning Framework

Reinforcement learning (RL) is used as one of the techniques to learn optimal control for autonomous agents in unknown environment [12]. The main idea is that agent execute action a_t in particular state s_t, goes to the next state s_{t+1} and receives numerical reward r_{t+1} as a feedback of recent action. Agent should explore state space and for every state find actions, which is more rewarded than other in some finite horizon.

Let, $Q(s, a)$ – is a Q-function reflects quality of selecting specified action a in state s. The initial values of Q-function are unknown and equal to zero. The learning goal is to approximate optimal Q-function, e.g. finding true Q-values for each action in every state using received sequences of rewards during state transitions. Using Q-learning rule [12], the temporal difference error calculated by:

$$\delta^t = r^t - \gamma \max_{a \in A(s^{t+1})} Q(s^{t+1}, a) - Q(s^t, a^t) \tag{2}$$

Where r_t – reward value obtained for action α_t selected in s_t, and γ – discount rate, $A(s_{t+1})$ – set of actions available at s_{t+1}.

4.2   RL-model for Module Positioning

Using defined above Q-learning rule define more precisely RL-model for module positioning including state, action and reward function description. It can be formulated as learning to find such a behavior policy that minimizes φ_{err}.

Let, a *state* of agent will be pair of values $s^t = [\varphi_{err}{}^t, \omega^t]$. Action set $A_\omega = \{\emptyset, \omega_+, \omega_-\}$ is represented by value of angular speed from Table 2. Action of robot $a^t \in A_\omega$ is a change of angular speed $\Delta \omega_t$ for given moment of time t.

The learning system is given a positive *reward* when the robot orientation closer to the goal orientation ($\varphi_{err}{}^t \to 0$) using optimal speed ω_{opt} and a penalty when the orientation of the robot deviates from the correct or selected action does not optimal for the given position. The value of the reward is defined as:

$$r^t = R(\varphi_{err}^{t-1}, \omega^{t-1}) \tag{3}$$

Where R – is reward function which is represented by decision tree depicting in the fig 5.

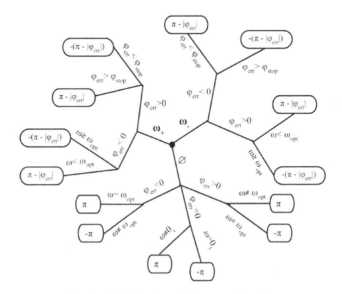

Fig. 5. A decision tree of the reward function

Here φ_{stop} – the value of angle, where robot reduce speed to stop at the correct orientation, ω_{opt} [0.6 .. 0.8] rad/s – optimal speed minimizing module power consumption. The parameter φ_{stop} is used to decrease the search space for the agent. When the agent angle error becomes smaller than φ_{stop}, an action that reduces the speed will get the highest award. The parameter ω_{opt} shows a possibility of power optimization by setting a value function. If agent angle error more than φ_{stop} and $\omega_{opt}^{min} < \omega < \omega_{opt}^{max}$, then agent award will get better increasing coefficient. This coefficient ranges between [0 .. 1]. The optimization gives the possibility to use preferred speed with the lowest power consumption.

5   Cooperative Moving

In this section, we consider a multi-agent reinforcement learning model for cooperative moving problem. The problem is to control module's individual speed in order to achieve stable circular motion of whole platform. Modules with different distances to beacon should have a different speed: for two modules i and j, with distances to beacon ρ_i and ρ_j respectively, the speed v_j will more than v_i if the distance to beacon ρ_j more than ρ_i. Every module should have additional policy to control its forward speed with respect to speed of other modules.

5.1   Multi-agent Reinforcement Learning Framework

The main principles of the technique are described in [13]–[14]. The basic idea of selected approach is to use influences between module and platform virtual leader to determine sequences of correct actions in order to coordinate behavior among them. The good influences should be rewarded and negative should be punished. The code design question is how to determine such influences in terms of received individual reward.

RL-framework used for such control problem is illustrated in Fig. 6:

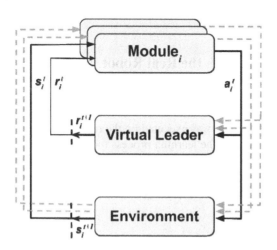

Fig. 6. Multi-Agent RL framework

The *i*-th module at the state sit selects action αit using current policy Qi and goes to next state s_i^{t+1} taking action to environment. Platform observes changes done by executed action, calculates and assigns reward r_i^{t+1} to module as a feedback reflecting successiveness of specified action.

The same *Q*-learning rule (2) can be used to update module control policy. The main difference between both rules is that in second case reward is assigned by a virtual leader instead of environment:

$$\Delta Q_i(s_i^t, a_i^t) = \alpha[r_{p \to i}^{t+1} + \gamma \max_{a \in A(s_i^{t+1})} Q_i(s_i^{t+1}, a) - Q_i(s_i^t, a_i^t)] \tag{4}$$

Instead of trying to build global Q-function Q($\{s_1, s_2, ..., s_n\}, \{a_1, a_2, ..., a_n\}$) for n modules we decompose the problem and build set of local Q-functions – $Q_1(s, a)$, $Q_2(s, a), ..., Q_N(s, a)$, where every policy contains specific control rule for each module.

The combination of such individual policies produces cooperative control law.

5.2   RL-model for Cooperative Moving

Let, state of the module is pair of $s_t = \{v_t, \overline{d}_i^{err} d_i^{err}\}$, where v_t – current value of linear speed, and $\overline{d}_{err}^t$ – distance error calculated by (4). Action set $A_v = \{\varnothing, v_+, v_-\}$ is represented by increasing/decreasing of linear speed from the Table 2 and action at∈ A_v is a change of forward speed Δv_t for given moment of time t.

The virtual agent receives error information for each module and calculates displacement error. This error can be positive (module ahead of the platform) or negative (the module behind of the platform). The learning process follows toward to minimization of $\overline{d}_i^{err}$ for every module. The maximum reward is given for case where $\overline{d}_i^{err} \to 0$, and a penalty given when the position of the module deviates from the predefined.

6   Experiments with the Real Robot

The simulation results showed in previous publication [4]. In this paper we consider on experiment with the real robot to make verification of the control system.

The learning of the agent was executed on the real robot after a simulation with the same external parameters. The learning process took 1440 iterations. The topology of Q-function is shown in Fig. 7. A real learning process took more iterations in average because the real system has noise and errors of sensors. Figure 8 illustrates the result of execution of a studied control system to turn modules to the center which is placed behind right.

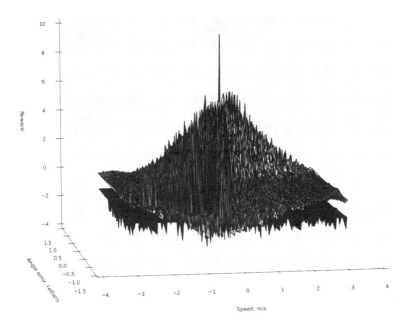

Fig. 7. Result topology of Q-function

Fig. 8. Execution of a learned control system to turn modules to the center that placed behind right relatively to the platform

Figure 9 shows the experimental result of the cooperative movement after learning positioning. The knowledge base of the learned agents was transferred to the agents of the control system on the real robot. Fig. 9 demonstrates the process of the platform moving by the learned system. At first, modules turn in the driving direction relative to the center of rotation (the circle drawn on white paper), as shown in screenshots 1-6 in Fig. 9. Then, the platform starts driving around the center of rotation in screenshots 7-9 in Fig. 9. The stabilization of the real module orientation is based on a low-level controller with feedback. This controller is provided by software control system of the robot. It helps to restrict the intellectual control system by manipulating linear speed of modules. As shown, the distance to the center of rotation is always the same on the entire trajectory of the platform.

Fig. 9. The experiment of modules turning to car kinematics scheme (1-6 screenshots) and movement around white beacon (7-9)

7   Conclusions

This paper provides implementation of an efficient, flexible, adaptive architecture for the control of a multi-wheeled production mobile robot. The system is based on decomposition into a holonic homogenous multi-agent system and on influence-based multi-agent reinforcement learning. The advantages of this method are follows:

- *Decomposition* means that the instead of trying to build global Q-function we build a set of local Q-functions.
- *Adaptability* – the platform will adapt its behavior for dynamically assigned beacon and will auto reconfigure moving trajectory.
- *Scalability* and *generalization* – the same learning technique is used for every agent, for every beacon position and every platform configuration.

The paper presents successful experiments with the real robot. Developed system provides robust steering of the platform for circular motion. The experiment results indicate that the application of the intellectual adaptive control system for real mobile robot have great prospects in a production.

In future works we will consider on comparison of the developed approach with existing approaches of a mobile robot steering and will provide further information about efficiency of the developed control system.

References

1. Andreas, J.C.: Energy-Efficient Electric Motors, 2nd edn. Marcel Dekker, New York (1992)
2. de Almeida, A.T., Bertoldi, P., Leonhard, W.: Energy efficiency improvements in electric motors and drives. Springer, Berlin (1997)
3. Stetter, R., Ziemniak, P., Paczynski, A.: Development, Realization and Control of a Mobile Robot. In: Obdržálek, D., Gottscheber, A. (eds.) EUROBOT 2010. CCIS, vol. 156, pp. 130–140. Springer, Heidelberg (2011)
4. Dziomin, U., Kabysh, A., Golovko, V., Stetter, R.: A multi-agent reinforcement learning approach for the efficient control of mobile robot. In: IEEE 7th International Conference on Intelligent Data Acquisition and Advanced Computing Systems, Berlin, vol. 2, pp. 867–873 (2013)
5. Mei, Y., Lu, Y.-H., Hu, Y.C., Lee, C.G.: Energy-efficient motion planning for mobile robots. In: 2004 IEEE International Conference on Robotics and Automation: Proceedings, ICRA 2004, vol. 5, pp. 4344–4349 (2004)
6. Ogunniyi, S., Tsoeu, M.S.: Q-learning based energy efficient path planning using weights. In: 24th Symposium of the Pattern Recognition Association of South Africa, pp. 76–82 (2013)
7. Mei, Y., Lu, Y.-H., Lee, C.G., Hu, Y.C.: Energy-efficient mobile robot exploration. In: IEEE International Conference, Robotics and Automation, pp. 505–511. IEEE Press (2006)
8. Ceccarelli, N., Di Marco, M., Garulli, A., Giannitrapani, A.: Collective circular motion of multi-vehicle systems with sensory limitations. In: 44th IEEE Conference, Decision and Control, 2005 and 2005 European Control Conference, pp. 740–745 (2005)
9. Ceccarelli, N., Di Marco, M., Garulli, A., Giannitrapani, A.: Collective circular motion of multi-vehicle systems. Automatica 44(12), 3025–3035 (2008)
10. Benedettelli, D., Ceccarelli, N., Garulli, A., Giannitrapani, A.: Experimental validation of collective circular motion for nonholonomic multi-vehicle systems. Robotics and Autonomous Systems 58(8), 1028–1036 (2010)
11. Ren, W., Sorensen, N.: Distributed coordination architecture for multi-robot formation control. Robotics and Autonomous Systems 56(4), 324–333 (2008)
12. Sutton, R.S., Barto, A.G.: Reinforcement learning: An introduction. MIT Press (1998)
13. Kabysh, A., Golovko, V.: General model for organizing interactions in multi-agent systems. International Journal of Computing 11(3), 224–233 (2012)
14. Kabysh, A., Golovko, V., Lipnickas, A.: Influence Learning for Multi-Agent Systems Based on Reinforcement Learning. International Journal of Computing 11(1), 39–44 (2012)

A Low-Cost Mobile Robot for Education

Valery Kasyanik and Sergey Potapchuk

Department of Intelligent Information Technology
Brest State Technical University
Moskovskaya 267, Brest 224017 Belarus
vvkasyanik@bstu.by
sergeipotapchuk@gmail.com

Abstract. We present, in this paper, a mobile robot and a program to control it that can be used in robotics education, expecting that the idea evolves, in a visible future, into a more general research tool in the field of robotics. The information necessary for robotic students to design their own mobile robot is, availability of each component of the robot, how each module of the robot is constructed, how these modules are combined, how a sensor system is given, how a simulator for the robot is programmed, what environment is appropriate to test the simulators, and so on. Also, it might be better if a possibility is given for developing it in student's home.

Keywords: mobile robot, education, low-cost robotics.

1   Introduction

Education of robotics is not easy. Students need knowledges from various scientific areas to carry out high quality researches. Robotics is one of the most important subjects and most active areas in IT education today. In fact, lots of robots for educational purpose have been proposed [1,2,3,4,5]. If these robots are available, then students may start a course simply by designing a tool for training these robots. On the other hand, if students start by creating an actual robot, they will learn a deeper methodology in robotics. In this paper, we consider both of such possibilities.

Unlike textbook problems, real world problems usually have multiple solutions. Moreover, it is the real world not us which decides whether our design or our hypothesis is correct or not. We could not ignore resource limitations such as time, money and materials. And more importantly, real world would less likely follow our models. For example, the assumption that sensors always deliver exact and valid values, or, that motors always deliver the commanded speed and torque will not work as planned due to noises, spurious inputs, unreliable outputs, etc. Students who lack hands-on experiences significantly underestimate the importance of these real world issues.

Hence our purpose is to develop a universal system for robotics for education that consists of constructing hardware of mobile robots, and software to control the robot in any environment given so that we can expect this system to work not only in a virtual world but also in the real world.

V. Golovko and A. Imada (Eds.): ICNNAI 2014, CCIS 440, pp. 182–190, 2014.
© Springer International Publishing Switzerland 2014

1.1   Related Works

We started this project along the ideas already reported [6,7]. The key features are: low cost, modularity, simplicity of programming and multi-functionality. Each of the above mentioned authors tried to find a balance between these qualities.

Also, a lot of interesting thought was found in Raymond's robot [8], although it was not for universal task but a particular task of rescue operations. A suggestion by Beer et al. [10] also attracts our interest. They used a platform based on LEGO [11]. However, LEGO is somehow a difficult platform for us to use together with the common packages available in public to control a robot.

The platform and prototype called 'Robot Pioneer 3DX available from AdeptRobotics [12] is one of the most attractive tools for us. It gives us a platform for two wheel mobile robot as well as a set of algorithms to control the robot by PC. The problem for us is its price. The idea of combining the virtual robot with a real robot was implemented by Gerkey et al. [13]. Their Player/Stage Project, tools for multi-robot and distributed sensor systems, is popular in the community of the mobile robot nowadays. It's often conveniently used together with commercial mobile robot of Pioneer 3-DX available from Adept Mobilerobots inc. [14].

We use a similar but more easily accessible mobile robot MARVIN (Mobile Autonomous Robot & Virtual Intelligent ageNt). Differences of our work from such already reported projects described above, and hopefully our novelties are that our mobile robots are our home made, partly because of the high price of the commercial products, and more importantly, in this way we have a possibility of more effective mobile robot and software for it than a commercial one. Furthermore, this allows us to make students study methods of artificial intelligence not only on models or by theory but also with experiments with using a physical world.

2   Our Robot

2.1   Requirements

Requirements of such robots are universality, modularity, availability of components and restriction about cost. We have to find a balance between these requirements. From these aspects, we face a couple of choices as follows.

Platform Choice: For example we have to choose a platform of how it moves - caterpillar, wheel, walking, or something else. Our choice is wheel robots because this is considered to be more universal [15]. This choice might be also practical due to its low cost, simplicity and easy control.

Structure Choice: we choose "Plug-and-play" modularity, which allows us to combine modules quite easily and gives us an almost endless variety of structure modification afterwards.

Software Choice: Our decision is, programming should be made for micro computer on board in the robot and then executes a command from a remote computer.

Sensor System Choice: Sensor system should be fixed for a specific task in advance. we have to options of choice. One is to use a small number of expensive sensors give a common flow of data that grows huge as time goes. The other is a large number of cheap sensors each of which issue a small information one by one.

A robot such as Marvin prefers the second option, and a low-cost mobile robot for education goes for our robot.

2.2   Platform

As mentioned above, we use the MARVIN Robot, and we chose DFRobot's two-wheeled platform with Arduino Board. See Fig. 1.

Fig. 1. DFrobot 2WD mobile platform

Its homepage [16] reads, "Arduino is an open source electronics prototyping platform based on flexible, easy-to-use hardware and software with the intention for artists, designers, hobbyists and anyone interested in creating interactive objects or environments."

The platform mounts two differential drives powered by high-quality high-speed motors and flexible rubber wheels are mounted on a high-strength aluminum body. The motors can be controlled by a variety of micro-controllers, which allows the robot turn even with almost zero radius. The platform can be improved by increasing modularity.

Fig. 2. MARVIN robot

In Fig.2, we show such an example with modules: power supply, onboard computer, sensors and actuators. The location of each module is flexible. We put onboard microprocessor vertically between motors and sensors, which reduces the influence of hindrances from motors to the sensors. It also makes communication between modules simple and flexible.

The overall specification is, size of body 17 x 23 cm, weight 1.4 kg, wheel diameter 65 mm and maximum speed 61 cm/sec.

2.3   Software

The robot consists of three modules, that is, onboard computer, motor drivers and power subsystem. On board computer is also based on the Arduino's project. It operates all peripheral devices of the robot. See Fig. 5. A number of resources for the onboard computer, including sample programs, are freely available in the Internet. Therefore, our onboard computer as well as motor driver is made by ourselves in our laboratory.

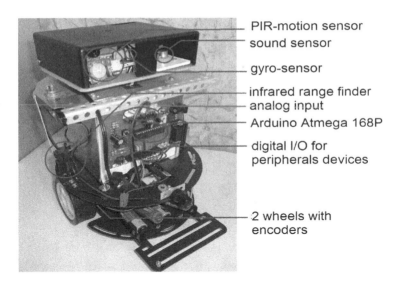

Fig. 3. Placement of sensors and onboard computer

For communication between the onboard computer and external devices, we can use wireless modules such as Bluetooth, Wi-Fi or connection by RS-232, which simplifies algorithms on the remote computer. We can also use a smartphone as a remote computer. Alternatively we could even replace the onboard computer with a smartphone.

2.4   Sensors

Quality of robot behavior strongly depends on the amount of information from sensors. The mobile robot MARVIN is equipped with various types of sensors: three infrared range finders, motion sensor, sound sensor, gyroscope, vibration sensor, temperature sensor, odometers, and touch sensors on its bumper.

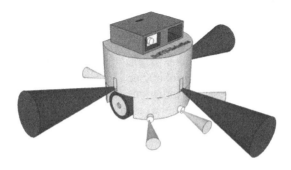

Fig. 4. Areas of collisions detection and obstacles

See Fig. 3. The sensor subsystem is also modular. It consists of two types of sensors: sensors for perception of the world outside and sensors of states of inside the robot. Locations of such sensors are shown in Fig. 6.

2.5   Environment

Our robot explores 'Webots'. Webots is a development environment used to model, program and simulate mobile robots. With Webots the user can design complex robotic setups, with one or several, similar or different robots, in a shared environment. See its web-site [17].

3   Programming

Under the Arduino platform, the program is usually fed to the robot through USB port when the robot is switched off. However we sometimes want a wireless access to the robot in order for the remote PC to be able to send an operating command on line, so that we can run algorithms on the remote PC in order for the robot to make actions just by receiving a series of commands. For the purpose we exploit the Firmata project. As its homepage [18] read, Firmata is a generic protocol for communicating with microcontrollers from software on a host computer. It is intended to work with any host computer software package. It is easy for other software to use this protocol. Basically, this is a protocol for talking to the Arduino from the host software. The aim is to control the Arduino completely from software on the host computer.

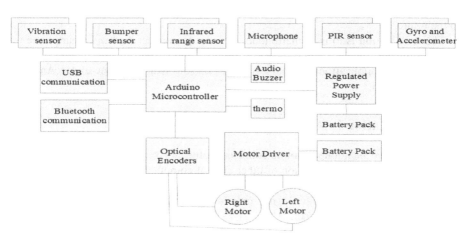

Fig. 5. Robot MARVIN functional diagram

To implement this communication, our first option is to use an open source software ROS (Robot Operating System) which is free for research use. ROS.org [19] says ROS is a framework for robot software development, providing an operating-system-like functionality such as low level device control, message passing between processes etc. ROS was originally developed in 2007 by the Stanford Artificial Intelligence Laboratory. As of 2008, development continues primarily at Willow Garage. ROS is framework based on ideas of Player/Stage and models from ROS can be used in Player/Stage simulator.

Development of the driver for our robot in webots simulator is now on going in our laboratory. See Fig. 6.

Fig. 6. Mobile autonomous robot and its 3D model

The second option is to use Matlab. Arduino project developed a firmware to enable communication with Matlab, and MARVIN supports this protocol [20]. At this moment the firmware can process data from sensors only via neural networks, but the project plan to provide firmwares for different peripheries other than neural networks in the future.

The third option is to use smartphone. In order to connect Android smartphone to the Arduino, we can use an open public library called Androino [21]. We can also create original applications by ourselves to control Arduino robot under Android operating system.

4   Application in Artificial Intelligent Area

Designing above mentioned mobile robot is useful, per se, in IT education. But we can use these robots for education of other area of computer science/technology. One of such areas is artificial intelligence. Through these projects, students can study methods of artificial intelligence not only on models or by theory but also with experiments in a physical world.

We now think of possibilities of two such areas. One is controlling our mobile robots by human voice. The aim is to make the robot intelligently understand our command and intelligently behave according to our commands. The other is path planning. The robot should intelligently decide a path from the start point to the goal avoiding obstacles.

There have been lots of research works in these two areas. Let's name a few. In the area of intelligent communication between human and mobile robot, we have, for example, papers entitled "MORPHA: Communication and interaction with intelligent, anthropomorphic robot assistants" [22], "Multi-robot cooperation for human-robot communication" [23], "A human-like semi autonomous mobile security robot." [24], "Modeling human-robot interaction for intelligent mobile robotics." [25], "Socially intelligent robots: dimensions of human-robot interaction." [26], "Human-like interaction skills for the mobile communication robot robotinho." [27] and so on.

In the area of intelligent path planning of mobile robot, we have "Mobile Robot Path Planning Using Hybrid Genetic Algorithm and Traversability Vectors Method." [28], "Path planning of autonomous mobile robot." [29], "An intelligent algorithm for the path planning of autonomous mobile robot for dynamic environment." [30] and so on.

These papers claim 'intelligence' but it's not real human-like intelligence indeed. So this topic is further innovative research topic, too.

5   Conclusion and Further Work

In this paper we have described how we construct mobile robot in education with a limited fund. We propose the two-wheel mobile robot MARVIN with a Plug-and-play modularity based on Arduino project. Robot moves by onboard microprocessor communicating with remote PC under Firmata protocol. We also proposed to use Android smartphone using the open public library Androino.

With this mobile robot we give tasks of two artificial intelligent area, "intelligent path planning" and "intelligent communication between human and robot". We expect this topic is not only for students but our future innovative research work.

Acknowledgments. We are very grateful to Professor Akira Imada for his important notes and corrections which made our paper much better.

References

1. K-Team Corporation - Mobile Robotics, http://www.k-team.com
2. Festo Didactic GmbH & Co.KG: Technical Documentation of Robotino. Martin Williams, Denkendorf (2007)
3. Mondada, F., Bonani, M., Raemy, X., Pugh, J., Cianci, C., Klaptocz, A., Magnenat, S., Zufferey, J.C., Floreano, D., Martinoli, A.: The e-puck, a Robot Designed for Education in Engineering. In: 9th Conference on Autonomous Robot Systems and Competitions, pp. 59–65. IPCB: Instituto Politécnico de Castelo Branco, Portugal (2009)
4. Surveyor SRV-1 Blackfin Robot, http://www.surveyor.com/
5. Nourbakhsh, I.: Robotics and education in the classroom and in the museum: On the study of robots, and robots for study. In: Workshop for Personal Robotics for Education. IEEE ICRA (2000)
6. Kumar, D., Meeden, L.: A robot laboratory for teaching artificial intelligence. In: 29th SIGCSE Symposium on Computer Science Education, pp. 341–344. ACM, New York (1998)
7. Schilling, K., Roth, H., Rusch, O.: Mobile Mini-Robots for Engineering Education. Global Journal of Engineering Education 6, 79–84 (2002)
8. Raymond, S.: The Building of Redback. In: 2005 Rescue Robotics Camp, Istituto Superiore Antincendi, Rome (2005)

9. Beer, R., Chiel, H., Drushel, R.: Using autonomous robots to teach science and engineering. Communications of the ACM (1999)
10. LEGO Education 2012 (2012), http://www.legoeducation.com
11. Lego Education WeDo, http://www.legoeducation.us/eng/product/lego_education_wedo_robotics_construction_set/2096
12. Intelligent Mobile Robotic Platforms for Service Robots, Research and Rapid Prototyping, http://www.mobilerobots.com/Mobile_Robots.aspx
13. Gerkey, B., Vaughan, R., Howard, A.: The Player/Stage Project: Tools for Multi-Robot and Distributed Sensor Systems. In: 11th International Conference on Advanced Robotics, Coimbra, Portugal, pp. 317–323 (2003)
14. Software for Pioneer DX, http://www.mobilerobots.com/Software.aspx
15. DFrobot 2WD mobile platform, http://www.dfrobot.com
16. Webpage of Arduino Project, http://www.arduino.cc
17. Webots Simulator, http://www.cyberbotics.com/overview
18. Firmata Project, http://firmata.org
19. Robotics Operation System, http://www.ros.org
20. Arduino Support from MATLAB, http://www.mathworks.com/academia/arduino-software/arduino-matlab.html
21. Interfacing android and arduino through an audio connection, http://androino.blogspot.com/p/project-description.html
22. Lay, K., Rassler, E., Dillmann, R., Grunwald, G., Hagele, M., Lawitzky, G., Stopp, A., von Seelen, W.: MORPHA: Communication and interaction with intelligent, anthropomorphic robot assistants. In: The International Status Conference - Lead Projects Human-Computer-Interactions (2001)
23. Kanda, T., Ishiguro, H., Ono, T., Imai, M., Mase, K.: Multi-robot cooperation for human-robot communication. In: IEEE Int. Workshop on Robot and Human Communication (ROMAN 2002), pp. 271–276. IEEE Press, Berlin (2002)
24. Carnegie, D.A., Prakash, A., Chitty, C., Guy, B.: A human-like semi autonomous mobile security robot. In: 2nd International Conference on Autonomous Robots and Agents, Palmerston North (2004)
25. Rogers, T.E., Peng, J., Zein-Sabatto, S.: Modeling human-robot interaction for intelligent mobile robotics. In: IEEE International Workshop on Robot and Human Interactive Communication, pp. 36–41. IEEE Press (2005)
26. Dautenhahn, K.: Socially intelligent robots: dimensions of human-robot interaction. Philosophical Transactions of the Royal Society B: Biological Sciences 362(1480), 679–704 (2007)
27. Nieuwenhuisen, M., Behnke, S.: Human-like interaction skills for the mobile communication robot robotinho. International Journal of Social Robotics 5(4), 549–561 (2013)
28. Loo, C.-K., Rajeswari, M., Wong, E.K., Rao, M.B.C.: Mobile Robot Path Planning Using Hybrid Genetic Algorithm and Traversability Vectors Method. Journal of Intelligent Automation & Soft Computing 10(1), 51–63 (2004)
29. Hachour, O.: Path planning of autonomous mobile robot. International Journal of Systems Application, Engineering & Development 2(4), 178–190 (2008)
30. Sarkar, S., Shome, S.N., Nandy, S.: An intelligent algorithm for the path planning of autonomous mobile robot for dynamic environment. In: Vadakkepat, P., et al. (eds.) FIRA 2010. CCIS, vol. 103, pp. 202–209. Springer, Heidelberg (2010)

Data-Driven Method for High Level Rendering Pipeline Construction

Victor Krasnoproshin and Dzmitry Mazouka

Belarusian State University, Minsk, Belarus
krasnoproshin@bsu.by, mazovka@bk.ru

Abstract. The paper describes a software methodology for the graphics pipeline extension. It is argued that common modern visualization techniques do not satisfy current visualization software development requirements adequately enough. The proposed approach is based on specialized formal language called visualization algebra. By invoking data-driven design principles inherited from the existing programmable pipeline technology, the technique has a potential to reduce visualization software development costs and build a way for further computer graphics pipeline automation.

Keywords: graphics pipeline, visualization, visualization algebra.

1 Introduction

Computer graphics (CG) remains one of the most rich and constantly evolving fields of study in computer science. CG consists of two large parts, each studying its own problem: image recognition and image generation. Both of these problems are very broad and complex in nature. In this paper, we concentrate on image generation, i.e. visualization.

Literally every area in human life which involves computer technology requires some sort of visual representation of information. Accurate and adequate visualization becomes vitally necessary with the growing complexity of problems being solved with computers. CG provides tools and theories that target the growing requirements for visualization, but as requirements become more complex and demanding so does the need for improvement in this field.

In this paper we analyze the most widely used visualization methodology and provide a technical solution for its improvement.

1.1 Related Works

Graphics pipeline design has been studied closely by game development companies and graphics hardware manufacturers. In the closest related work – "Advanced Scenegraph Rendering Pipeline" (nVidia) [1], authors recognize deficiencies of the common visualization system architectural approach. They offer unified structured data lists which can be processed uniformly in a data-oriented way. This allows to

V. Golovko and A. Imada (Eds.): ICNNAI 2014, CCIS 440, pp. 191–200, 2014.

decrease redundant computations determined by the previous architecture. Developers of Frostbite rendering engine [2] distinguish three major rendering stages in their architecture: culling (gathering of visible entities), building (construction of high level states and commands) and rendering (flushing states and commands to graphics pipeline). These stages represent independent processing events and can run on different threads simultaneously.

In our work we do not concentrate on performance problems of visualization systems. Taking into account the growing recognition of high level data-oriented design principles in graphics development, we are trying to provide a base line for a general solution from the system design's point of view. Technical implementation (Objects Shaders) naturally emerges from the existing shader languages of the graphics pipeline [3].

1.2   Previous Works

In our previous works [4, 5] we studied a theoretical model of a generalized visualization system. We introduced mathematical constructs for objects data space and provided a formalized algebra operating in that space. This work is a practical implementation of those formal concepts.

2   Basic Definitions

Visualization, broadly defined, is a process of data representation in visual form. In computer graphics visualization has a special definition – **rendering**. The rendering process is implemented in computers with a set of dedicated hardware components and specialized software.

There are several rendering algorithms that lay a base for computer visualization. The most widespread are ray tracing and **rasterization** [6]. These algorithms have their own application areas: ray tracing is used for photorealistic rendering, sacrificing computation performance to physically correct output; rasterization, in its turn, is designed for high-performance dynamic applications where photorealism is not as essential.

Almost all contemporary rendering hardware implements the rasterization algorithm. A technical implementation of the algorithm is called a **graphics pipeline**. A graphics pipeline is structured as a sequence of data processing stages, most of which are programmable. These stages are responsible for data transformations specified by the rasterization algorithm's logic. The pipeline is accessible for software developers with special hardware abstraction layer libraries, available on most operating systems. The most popular libraries are DirectX (for Windows) and OpenGL (for any OS). Both of them provide a software abstraction of the underlying hardware implementation of graphics pipeline with all necessary access methods.

The graphics pipeline is a very efficient and flexible technology, but it may be difficult to use in complex applications. This is because the pipeline's instructions operate on a low level with objects like geometrical points and triangles. The best analogy here would be to compare this with the efforts of programming in an assembly language.

The problem of the pipeline's complexity gave rise to a new class of visualization software: **graphics engines**. A graphics engine is a high level interface that wraps around the pipeline's functionality, and introduces a set of tools which are much more convenient to use in real world applications. The engine is normally built around an extendible but nevertheless static computer model, which generalizes a whole spectrum of potential visualization tasks. The most popular visualization model for graphics engines nowadays is the **scene graph**.

Scene graph is not a well defined standard model and many software developers implement it differently for different tasks. However, the implementations often have common traits, which can be summarized in the following definition. Scene graph – is a data structure for visualization algorithms, based on a hierarchical tree where every node is an object and every subnode – a part of that object. That is, a scene graph may have a node representing a chair object and direct subnodes representing its legs. Furthermore, objects may be assigned with so-called materials – fixed properties lists describing how particular objects should be rendered.

A primitive scene graph's rendering algorithm consists of visiting every node of the tree and rendering each of the objects individually using material properties.

3   Analysis

The following figure (Fig.1) summarizes the structure of the visualization process and differences between two approaches mentioned above.

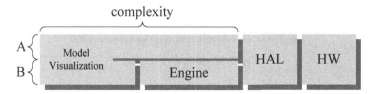

Fig. 1. Visualization process

HW and HAL stand correspondingly for graphics hardware and hardware abstraction layer represented by DirectX or OpenGL. Engine stage is the part of the process taken over by a graphics engine.

The rest of the process is marked as model visualization. This stage is directly connected to the very visualization problem in question. Computer graphics, like any other computing activity, works with **computer models**. These models may represent any kinds of real or imaginary entities, phenomena or processes. Model visualization is essentially a set of algorithms translating the model into another form of representation which is suitable for automatic rendering.

Earlier, we described two common visualization approaches: pure graphics pipeline rendering and graphics engines. They are marked on the scheme as A and B respectively. It is emphasized how graphics engines cut off a serious portion of implementation complexity. However, the real situation is not that simple.

We have mentioned that graphics engines is a whole class of non-standard software. And, moreover, this software is built on the base of certain fixed assumptions, and implements fixed input models that aren't necessarily compatible with the current target model. This situation is depicted on the next figure (Fig.2).

Fig. 2. Model and engines

The model on the picture is not compatible with either of two engines. This happens, for example, in the case when the engines specialize in solid 3D geometric object rendering, and the Model represents a flat user interface.

In this situation software engineers can take one of the three ways:

1. change the model so it fits an engine;
2. provide an engine-model adapter layer;
3. fallback to the pure pipeline rendering.

The first option probably needs the least effort, but the outcome of visualization may seriously differ from the initial expectations as the visualized model gets distorted.

The second option tries to preserve the model's consistency with additional translation stage (Fig.3):

Fig. 3. Model-engine adapter

This may work in some situations depending on how different the model and engine are. If the difference is too big, the adapter itself becomes cumbersome and unmaintainable. In this case the third option becomes preferable: the visualization problem gets solved from the scratch.

The conclusion of this is that we cannot rely on graphics engines from a general perspective. Tasks and models change all the time and engines become obsolete. The variety and quantities grow, which make it troublesome to find the proper match. And so, we have a question: whether anything can be done here in order to improve the visualization process construction.

In our previous works [4, 5] we suggested that it was possible to develop a general methodology for high-level visualization abstractions. That is, to create a model-independent language for visualization process construction.

Together with corresponding support layer libraries, the new visualization process construction would change in the following way (Fig.4):

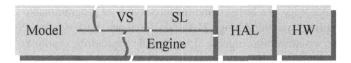

Fig. 4. A new way of the process construction

In this scheme, the model is rendered using an engine with either the first or the second method. SL stands for the standard layer, and VS – for visualization system.

Standard layer is a set of support rendering libraries based on visualization algebra methodology [5]. The library provides tools for process construction in model-agnostic data-driven way. Less effort (in comparison to pure pipeline development) is necessary to make a model adaptation visualization system. And, though the implementation complexity remains slightly bigger than that of the graphics engines, this technique has the advantage of preserving the pipeline's flexibility together with higher level of language abstraction.

4   Object Shaders

Before we start describing the details of the proposed technology, a few words needs to be said regarding how the graphics pipeline is programmed in general.

The graphics pipeline consists of several stages including (DirectX11 model [7]): Input Assembler, Vertex Shader, Hull Shader, Tesselator, Domain Shader, Geometry Shader, Rasterizer, Pixel Shader and Output Merger.

The stages implement different parts of the rasterization algorithm and provide some additional functionality. Shader stages are programmable; all the others are configurable. Configurable stages implement fixed algorithms with some adjustable parameters. Programmable stages, in their turn, can be loaded with custom programs (which are called shaders) implementing specific custom algorithms. This is what essentially gives the pipeline its flexibility.

Shader programs of different types operate with different kinds of objects. Vertex shaders operate with geometry vertices and control geometrical transformation of 3D model and its projection onto 2D screen. Pixel shaders work with fragments – pieces of the resulting image that are later translated into pixels.

The most common way in writing the shader programs is to use one of the high level shading languages: HLSL for DirectX, or GLSL for OpenGL. These languages have common notation and similar instruction sets determined by underlying hardware. By their nature, the languages are vector processing SIMD-based. And corresponding shader programs implement algorithms that describe how every vertex or fragment needs to be treated independently, enabling massive parallel data processing capability.

In our work, we pursue technological integration into the graphics pipeline, rather than its replacement with another set of tools. That is why the technical realization of the proposed methodology is based on emulation of an additional programmable pipeline stage which we call Object Shader.

Object Shader stage is a broad name for three types of programmable components: pipeline, sampling, and rendering. These components are based on corresponding notions from visualization algebra [5]: visual expression, sample, and render operations.

Visual expression in visualization algebra (VA) is a formalized algorithm operating in a generalized object space. Objects in VA are represented with tuples of attributes projected onto a model-specific semantic grid. The methodology does not make any assumptions on the nature of objects and their content, treating all data equally.

Visual expressions in VA are constructed using four basic operations:

1. sample – object subset selection,
2. transform – derivation of new objects on the base of existing ones,
3. render – objects translation into an image,
4. blend – operations with the resulting images.

The final expression for target model visualization must have one input (all the model's data) and one output (the resulting image).

On the technical side, the expression is represented with a program on a language similar to HLSL. The additions are:

1. data type ptr used for declaration of resources
2. object types Scene, Objects and Frame
3. various rendering-specific functions

Object layout declaration in object shaders is done in a common way with structures:

```
struct Object
{
  field-type field-name : field-semantic;
  ...
};
```

Sampler functions are simple routines returning boolean values:

```
bool Sample(Object object)
{
  return sampling-condition;
}
```

Render procedures generate operation sequences for objects of the supported type and return Frame as a result:

```
Frame Render(Object object)
{
  instruction;
  ...
  instruction;
  return draw-instruction;
}
```

Pipeline procedures combine sampling and rendering operations into the final visualization algorithm. The input of a pipeline procedure is a Scene object and a Frame object is output:

```
Frame Pipeline(Scene scene)
{
  Objects list = Sample(scene);
  ...
  Frame frame = Render(list);
  return frame;
}
```

Visualization system routes data streams according to the pipeline procedure logic, splitting it with samplers and processing with renderers. The unit routines are designed to be atomic in the same way how it is done for the other shader types: processing one object at a time, allowing massive parallelization.

This technique is fairly similar to effects in DirectX [8]; but, at the same time, the differences are apparent: effects cannot be used for building the complete visualization pipeline.

In the last part of the paper we will go through the real usage example.

5   Usage Example

The sample model consists of one 3D object (a building) and one 2D object (overlay image). The resulting visualization should visualize the building at the center of screen and make it possible to change its orientation. The overlay image should be drawn at the top right corner.

From the model description we know that there are two types of objects and therefore we need two sampling and two rendering procedures.
Sampling procedures:

```
bool Sample1(int type : iType)
{
  return type == 1;
}

bool Sample2(int type : iType)
{
  return type == 2;
}
```

The procedures get the type field from object stream and compare it against predefined constant values. So the first procedure will sample 3D objects (type 1) and the second 2D objects (type 2).

Rendering procedure for 3D objects starts with object layout declaration:

```
struct Object
{
  float4x4 transform : f4x4Transform;
  ptr VB : pVB;
  ptr VD : pVD;
  ptr tx0 : pTX0;
  int primitive_count : iPrimitiveCount;
  int vertex_size : iVertexSize;
};
```

The supported objects must have transform matrix, vertex buffer (VB), vertex declaration (VD), texture (tx0) and common geometry information: primitives count and vertex size.

Then, we declare external variables, which are required to be provided by the user:

```
extern ptr VS = VertexShader("/Test#VS_World");
extern ptr PS = PixelShader("/Test#PS_Tex");
extern float4x4 f4x4ViewProjection;
```

These variables are: common vertex shader (VS), common pixel shader (PS) and view-projection transformation matrix.

The rendering procedure itself:

```
Frame OS_Basic(Object obj)
{
  SetStreamSource(0, obj.VB, 0, obj.vertex_size);
  SetVertexDeclaration(obj.VD);
  SetVertexShader(VS);
  SetPixelShader(PS);
  SetVertexShaderConstantF(0, &obj.transform, 4);
  SetVertexShaderConstantF(4, &f4x4ViewProjection, 4);
  SetTexture(0, obj.tx0);
  return DrawPrimitive(4, 0, obj.primitive_count);
}
```

The procedure makes a number of state change calls and invokes a drawing routine.

The second rendering procedure is implemented in a similar way:

```
struct Object
{
  float4 rect : f4Rectangle;
  ptr tx0 : pTX0;
};


extern ptr VB = VertexBuffer("/Test#VB_Quad");
```

```
extern ptr VD = VertexDeclaration("/Test#VD_P2");
extern ptr VS = VertexShader("/Test#VS_Rect");
extern ptr PS = PixelShader("/Test#PS_Tex");
Frame OS_UI(Object obj)
{
  SetStreamSource(0, VB, 0, 8);
  SetVertexDeclaration(VD);
  SetVertexShader(VS);
  SetPixelShader(PS);
  SetVertexShaderConstantF(0, &obj.rect, 1);
  SetTexture(0, obj.tx0);
  return DrawPrimitive(4, 0, 2);
}
```

The resulting pipeline procedure is very simple: it needs to use the declared samplers and renderers, and combine their output:

```
Frame PP_Main(Scene scene)
{
  return OS_Basic(Sample1(scene))+OS_UI(Sample2(scene));
}
```

Model data in Json format:

```
"Object1" :
{
  "iType" : "int(1)",
  "f4x4Transform" : "float4x4(0.1,0,0,0,  0,0.1,0,0,
    0,0,0.1,0,  0,-1,0,1)",
  "pVB" : "ptr(VertexBuffer(/Model#VB_House))",
  "pVD" : "ptr(VertexDeclaration(/Model#VD_P3N3T2T2))",
  "pTX0" : "ptr(Texture2D(/Model#TX_House))",
  "iPrimitiveCount" : "int(674)",
  "iVertexSize" : "int(40)"
},
"Rect1" :
{
  "iType" : "int(2)",
  "pTX0" : "ptr(Texture2D(/Test#TX_Tex))",
  "f4Rectangle" : "float4(0.75, 0.75, 0.25, 0.25)"
}
```

And the following picture represents the result of visualization (Fig.5):

Fig. 5. Result of visualization

6   Conclusion

Computer visualization still holds the status of a heavily evolving scientific and engineering area. Dozens of new techniques and hardware emerge every year. And, with further advancements, this environment may require certain intensive changes in order to stay comprehensible and maintainable.

This paper provides justification and a short overview of the technological implementation of the new visualization methodology based on so-called visualization algebra. This methodology has a potential to improve the most popular existing methods of visualization in terms of flexibility and accessibility.

References

1. Tavenrath, M., Kubisch, C.: Advanced Scenegraph Rendering Pipeline. In: GPU Technology Conference, San Jose (2013)
2. Andersson, J., Tartarchuk, N.: Frostbite Rendering Architecture and Real-time Procedural Shading Texturing Techniques. In: Game Developers Conference, San Francisco (2007)
3. MSDN, Programming Guide for HLSL, http://msdn.microsoft.com/en-us/library/windows/desktop/bb509635%28v=vs.85%29.aspx
4. Krasnoproshin, V., Mazouka, D.: Graphics pipeline automation based on visualization algebra. In: 11th International Conference on Pattern Recognition and Information Processing, Minsk (2011)
5. Krasnoproshin, V., Mazouka, D.: Novel Approach to Dynamic Models Visualization. Journal of Computational Optimization in Economics and Finance 4(2-3), 113–124 (2013)
6. Gomes, J., Velho, L., Sousa, M.C.: Computer Graphics: Theory and Practice. A K Peters/CRC Press (2012)
7. MSDN, Graphics Pipeline, http://msdn.microsoft.com/en-us/library/windows/desktop/ff476882%28v=vs.85%29.aspx
8. MSDN, Effects (Direct3D 11), http://msdn.microsoft.com/en-us/library/windows/desktop/ff476136%28v=vs.85%29.aspx

Efficiency of Parallel Large-Scale Two-Layered MLP Training on Many-Core System

Volodymyr Turchenko and Anatoly Sachenko

Research Institute for Intelligent Computer Systems
Ternopil National Economic University
3 Peremogy Square, Ternopil, Ukraine
{vtu,as}@tneu.edu.ua

Abstract. The development of parallel batch pattern back propagation training algorithm of multilayer perceptron with two hidden layers and its parallelization efficiency research on many-core high performance computing system are presented in this paper. The model of multilayer perceptron and the batch pattern training algorithm are theoretically described. The algorithmic description of the parallel batch pattern training method is presented. Our results show high parallelization efficiency of the developed training algorithm on large scale data classification task on many-core parallel computing system with 48 CPUs using MPI technology.

Keywords: Parallel batch pattern training, multilayer perceptron, parallelization efficiency, data classification.

1   Introduction

Artificial neural networks (NNs) represent a very good alternative to traditional methods for solving complex problems in many fields, including image processing, predictions, pattern recognition, robotics, optimization, etc [1]. However, NN models require high computational load in the training phase (on a range from several hours to several days) depending on the problem. This is, indeed, the main obstacle to face for an efficient use of NNs in real-world applications. The development of parallel algorithms to speed up the training phase of NNs is one of the ways to outperform this obstacle. Due to the last technological achievements, many-core high performance computing systems have a widespread use now in research and industry. Therefore the estimation of the parallelization efficiency of such parallel algorithms on many-core high performance systems is an actual research task.

Taking into account the parallel nature of NNs, many researchers have already focused their attention on NNs parallelization. The authors of [2] investigate parallel training of multi-layer perceptron (MLP) on SMP computer, cluster and computational grid using MPI (Message Passing Interface) parallelization. The development of parallel training algorithm of Elman's simple recurrent neural network based on Extended Kalman Filter on multicore processor and Graphic

V. Golovko and A. Imada (Eds.): ICNNAI 2014, CCIS 440, pp. 201–210, 2014.

Processing Unit is presented in [3]. The authors of [4] have presented the development of parallel training algorithm of fully connected RNN based on linear reward penalty correction scheme. In our previous works, within the development of the parallel grid-aware library for neural networks training [5], we have developed the batch pattern back propagation (BP) training algorithm for multi-layer perceptron with one hidden layer [6], recurrent neural network [7], recirculation neural network [8] and neural network with radial-basis activation function [9] and showed their good parallelization efficiency on different high performance computing systems.

However the analysis of the state-of-the-art has showed that the parallelization of the MLP with two hidden layers of neurons was not properly investigated yet. For example, the authors of [2] have parallelized the MLP architecture 16-10-10-1 (16 neurons in the input layer, two hidden layers with 10 neurons in each layer and one output neuron) on the huge number of the training patterns (around 20000) coming from Large Hadron Collider. Their implementation of this relatively small NN with 270 internal connections (number of weights of neurons and their thresholds) does not provide positive parallelization speedup due to large communication overhead, i.e. the speedup is less than 1. In our opinion this overhead is caused by the fact, that the "communication part" of their algorithm is not optimized and contains at least three separate communication messages.

According to the theorem of universal approximation [10, 11], the MLP with two hidden layers provides better control on approximation process instead the MLP with one hidden layer [1]. Also the MLP with two hidden layers allows processing the "global features", i.e. the neurons of the first hidden layer gather the "local features" from input data and the neurons of the second hidden layer generalized their outputs providing higher-level representation. This is especially urgent for solving object classification, recognition and computer vision tasks. However the computational time for such kind of tasks is extremely long. For example, the training of the network with 500 neurons in the first hidden layer and 300 neurons in the second hidden layer on 60,000 training vectors of the MNIST database [12] was extremely slow; only 59 epochs of pre-training of the hidden layers as a Restricted Boltzmann Machine took about a week [13]. Therefore the research of efficiency of a parallel training algorithm for MLP with two hidden layers of neurons is an actual research problem.

Taking into account that the batch pattern BP training scheme showed good parallelization efficiency on the number of NN architectures, the goal of this paper is to apply this scheme for the MLP with two hidden layers and investigate its parallelization efficiency on the large-scale data classification task on a many-core high performance computing system. The rest of the paper is organized as follows: Section 2 details the mathematical description of batch pattern BP algorithm, Sections 3 describes its parallel implementation, Section 4 presents the obtained experimental results and Section 5 concludes this paper.

2   Batch Pattern BP Training Algorithm for Multilayer Perceptron with Two Hidden Layers

The batch pattern training algorithm updates neurons' weights and thresholds at the end of each training epoch, i.e. after all training patterns processing, instead of updating weights and thresholds after processing of each pattern in the sequential training mode [6]. The output value of the MLP with two hidden layers (Fig. 1) is described by [1]:

$$y_3 = F_3(S_3), S_3 = \sum_{k=1}^{K} y_{2k} \cdot w_{3k} - T_3, \tag{1}$$

$$y_{2k} = F_2(S_{2k}), S_{2k} = \sum_{j=1}^{N} y_{1j} \cdot w_{2jk} - T_{2k}, \tag{2}$$

$$y_{1j} = F_1(S_{1j}), S_{1j} = \sum_{i=1}^{M} x_i \cdot w_{1ij} - T_{1j}, \tag{3}$$

where F_3, F_2 and F_1 are the logistic activation functions $F(x) = 1/(1+e^{-x})$ which used for the neurons of the output and two hidden layers respectively, S_3, S_{2k} and S_{1j} are the weighed sums of the neurons of the output and two hidden layers respectively, K, N and M are the number of the neurons of two hidden and input layers respectively, y_{2k} are the outputs of the neurons of the second hidden layer, w_{3k} are the weight coefficients from the k-neuron of the second hidden layer to the output neuron, T_3 is the threshold of the output neuron, y_{1j} are the outputs of the neurons of the first hidden layer, w_{2jk} are the weights from the neurons of the first hidden layer to the neurons of the second hidden layer, T_{2k} are the thresholds of the neurons of the second hidden layer, x_i are the input values, w_{1ij} are the weights from the input neurons to the neurons of the first hidden layer and T_{1j} are the thresholds of the neurons of the first hidden layer.

The batch pattern BP training algorithm for this MLP architecture consists of the following steps [14]:

1. Set the desired Sum Squared Error SSE= E_{min} and the number of training epochs t;
2. Initialize the weights and thresholds of the neurons with values in range (0...0.5) [14];
3. For the training pattern pt:
    3.1. Calculate the output value $y_3^{pt}(t)$ by expressions (1), (2) and (3);
    3.2. Calculate the error of the output neuron $\gamma_3^{pt}(t) = y_3^{pt}(t) - d^{pt}(t)$, where $y_3^{pt}(t)$ is the output value of the MLP and $d^{pt}(t)$ is the target output value;

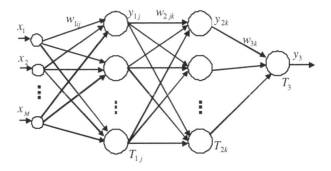

Fig. 1. The structure of a MLP with two hidden layers

3.3. Calculate the errors of the second hidden layer neurons'
$\gamma_{2k}^{pt}(t) = \gamma_{3}^{pt}(t) \cdot w_{3k}(t) \cdot F_{3}'(S_{3}^{pt}(t))$, where $S_{3}^{pt}(t)$ is the weighted sum of the output neuron and the errors of the first hidden layer neurons'
$\gamma_{1j}^{pt}(t) = \sum_{k=1}^{K} \gamma_{2k}^{pt}(t) \cdot w_{2jk}(t) \cdot F_{2}'(S_{2k}^{pt}(t))$, where $S_{2k}^{pt}(t)$ are the weighted sums of the k-neuron of the second hidden layer;

3.4. Calculate the delta weights and delta thresholds of all neurons and add the result to the value of the previous pattern

$$s\Delta w_{3k} = s\Delta w_{3k} + \gamma_{3}^{pt}(t) \cdot F_{3}'(S_{3}^{pt}(t)) \cdot y_{2k}^{pt}(t) , \quad s\Delta T_{3} = s\Delta T_{3} + \gamma_{3}^{pt}(t) \cdot F_{3}'(S_{3}^{pt}(t)), \quad (4)$$

$$s\Delta w_{2jk} = s\Delta w_{2jk} + \gamma_{2k}^{pt}(t) \cdot F_{2}'(S_{2k}^{pt}(t)) \cdot y_{1j}^{pt}(t) , \quad s\Delta T_{2j} = s\Delta T_{2j} + \gamma_{2k}^{pt}(t) \cdot F_{2}'(S_{2k}^{pt}(t)) , \quad (5)$$

$$s\Delta w_{1ij} = s\Delta w_{1ij} + \gamma_{1j}^{pt}(t) \cdot F_{1}'(S_{1j}^{pt}(t)) \cdot x_{i}^{pt}(t) , \quad s\Delta T_{1j} = s\Delta T_{1j} + \gamma_{1j}^{pt}(t) \cdot F_{1}'(S_{1j}^{pt}(t)) ; \quad (6)$$

3.5. Calculate the SSE using $E^{pt}(t) = \frac{1}{2}\left(y_{3}^{pt}(t) - d^{pt}(t)\right)^{2}$; $\qquad$ (7)

4. Repeat the step 3 above for each training pattern pt, where $pt \in \{1,...,PT\}$, PT is the size of the training set;

5. Update the weights and thresholds of neurons using the expressions
$w_{3k}(PT) = w_{3k}(0) - \alpha_{3}(t) \cdot s\Delta w_{3k}$, $\qquad\qquad T_{3}(PT) = T_{3}(0) + \alpha_{3}(t) \cdot s\Delta T_{3}$,
$w_{2jk}(PT) = w_{2jk}(0) - \alpha_{2}(t) \cdot s\Delta w_{2jk}$, $\qquad T_{2k}(PT) = T_{2k}(0) + \alpha_{2}(t) \cdot s\Delta T_{2k}$,
$w_{1ij}(PT) = w_{1ij}(0) - \alpha_{1}(t) \cdot s\Delta w_{1ij}$, $\quad T_{1j}(PT) = T_{1j}(0) + \alpha_{1}(t) \cdot s\Delta T_{1j}$ where $\alpha_{3}(t)$, $\alpha_{2}(t)$ and $\alpha_{1}(t)$ are the learning rates;

6. Calculate the total SSE $E(t)$ on the training epoch t using $E(t) = \sum_{pt=1}^{PT} E^{pt}(t)$;

7. If $E(t)$ is greater than the desired error E_{min} then increase the number of training epochs to $t+1$ and go to step 3, otherwise stop the training process.

3   Parallel Batch Pattern BP Training Algorithm of MLP with Two Hidden Layers

Similarly to the MLP with one hidden layer [6], the sequential execution of steps 3.1-3.5 for all training patterns is parallelized, because the sum operations $s\Delta w_{3k}$, $s\Delta T_3$, $s\Delta w_{2jk}$, $s\Delta T_{2k}$, $s\Delta w_{1ij}$ and $s\Delta T_{1j}$ are independent of each other. The computational work is divided among the *Master* (executing assigning functions and calculations) and the *Workers* (executing only calculations) processors.

The algorithms for *Master* and *Workers* are depicted in Fig. 2. The *Master* starts with definition (i) the number of patterns *PT* in the training data set and (ii) the number of processors *p* used for the parallel executing of the algorithm. The *Master* divides all patterns in equal parts according to the number of *Workers* and assigns one part of patterns to itself. Then the *Master* sends to *Workers* numbers of appropriate patterns to train.

Each *Worker* executes the following operations for each pattern *pt*:
1. calculate steps 3.1-3.5 and 4, only for its assigned number of training patterns. The values of the partial sums of delta weights $s\Delta w_{3k}$, $s\Delta w_{2jk}$, $s\Delta w_{1ij}$ and delta thresholds $s\Delta T_3$, $s\Delta T_{2k}$, $s\Delta T_{1j}$ are calculated there;
2. calculate the partial SSE for its assigned number of training patterns.

After processing all assigned patterns, only one allreduce collective communication operation (it provides the summation as well) is executed automatically providing the synchronization with other processors by its internal implementation [15]. However from the algorithmic point of view it is showed as an independent operation in Fig. 2 before the operation of data reduce. Then the summarized values $s\Delta w_{3k}$, $s\Delta T_3$, $s\Delta w_{2jk}$, $s\Delta T_{2k}$, $s\Delta w_{1ij}$ and $s\Delta T_{1j}$ are sent to all processors working in parallel and placed into the local memory of each processor. Each processor uses these values for updating the weights and thresholds according to the step 5 of the algorithm in order to use them in the next iteration of the training algorithm. As the summarized value of $E(t)$ is also received as a result of the reducing operation, the *Master* decides whether to continue the training or not.

The software is developed using C programming language with the standard MPI functions. The parallel part of the algorithm starts with the call of *MPI_Init()* function. The *MPI_Allreduce()* function reduces the deltas of weights $s\Delta w_{3k}$, $s\Delta w_{2jk}$, $s\Delta w_{1ij}$ and thresholds $s\Delta T_3$, $s\Delta T_{2k}$, $s\Delta T_{1j}$, summarizes them and sends them back to all processors in the group. Since the weights and thresholds are physically located in the different matrixes of the routine, we did pre-encoding of all data into one communication message before sending and reverse post-decoding the data to the appropriate matrixes after receiving the message in order to provide only one physical call of the function *MPI_Allreduce()* in the communication section of the algorithm. Function *MPI_Finalize()* finishes the parallel part of the algorithm.

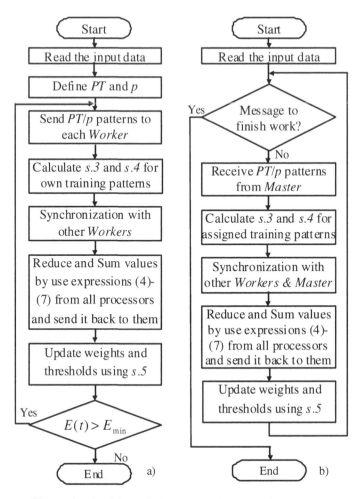

Fig. 2. The algorithms of Master (a) and Workers (b) processors

4   Experimental Results

The many-core high performance computing system *Remus*, located in the Innovative Computing Lab, the University of Tennessee, USA is used for the research. *Remus* consists of two socket G34 motherboards RD890 (AMD 890FX chipset) connected each other by AMD Hyper Transport. Each motherboard contains two twelve-core AMD Opteron 6180 SE processors with a clock rate of 2500 MHz and 132 GB of local RAM. Thus the total number of computational cores is 48. Each processor has the L2 cache of 12x512 Kb and the L3 cache of 2x6 Mb. We run the experiments using MPI library Open MPI 1.6.3 [16].

Analogously to the work [13], the MNIST database of handwritten digits, which contains 60,000 training images and 10,000 test images, is used for the research of the parallelization efficiency of large-scale data classification task. The size of the input images in the MNIST database is equal to the 784 elements; therefore the number of the input neurons of our MLP will be 784. In order to assess the different sizes of the computational problem, the number of the neurons of the first hidden layer was changed as 80, 160, 240, 320 and 400. Similarly, the number of the neurons of the second hidden layer was changed as 50, 100, 150, 200 and 250. There are ten classes in the database, the digits 0...9, thus our large-scale MLP model has only one output neuron. We did the training using continuously growing number of training images, from 10,000 patterns in the smallest scenario to 50,000 patterns in the biggest scenario. Thus, the following parallelization scenarios were researched: 784-80-50-1/10,000 patterns, 784-160-100-1/20,000 patterns, 784-240-150-1/30,000 patterns, 784-320-200-1/40,000 patterns, and 784-400-250-1/50,000 patterns. The last scenario is slightly lesser from the scenario 784-500-300-1/60,000 mentioned in the paper [13].

Results of our previous researches have showed that the parallelization efficiency of the parallel batch pattern training algorithm does not depend on the number of training epochs [17] since the neurons' weights and thresholds are combining in the end of each epoch. Therefore, taking into account the possible huge computational time of the whole experiment, we have researched the parallelization efficiency of the parallel training algorithm for one hundred training epoch only. The learning rates are constant and equal $\alpha_1(t) = 0.6$, $\alpha_2(t) = 0.6$ and $\alpha_3(t) = 0.6$. The expressions $S=Ts/Tp$ and $E=S/p \times 100\%$ are used to calculate the speedup and efficiency of parallelization, where Ts is the time of sequential executing of the routine, Tp is the time of executing the parallel version of the same routine on p processors of a parallel system.

The parallelization efficiency of the parallel batch pattern training algorithm for the large-scale MLP with two hidden layers of neurons for one hundred training epoch and the average communication (messaging) overhead on 5 researched scenarios are presented in Fig. 3 and Fig. 4 respectively. The experimental results (Fig. 3) showed high parallelization efficiency of the MLP with two hidden layers of neurons: 96-98% on 48 cores for the three bigger parallelization scenarios and 72-78% on 48 cores for the two smaller parallelization scenarios. Normally, the communication overhead is increasing with increasing the number of the cores used for the parallelization (Fig. 4).

Thus, taking into account obtained results, the large-scale data classification task which, for the successful results, may require, for example, 10^4 training epochs for the MLP 784-400-250-1 with two hidden layers of neurons on the 50,000 training patterns of the MNIST database will be computed approximately 21 days by sequential routine and only 10 and half hours on 48 cores of many-core high performance computing system.

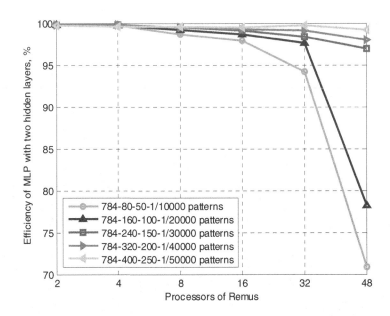

Fig. 3. Parallelization efficiency of the algorithm on many-core system

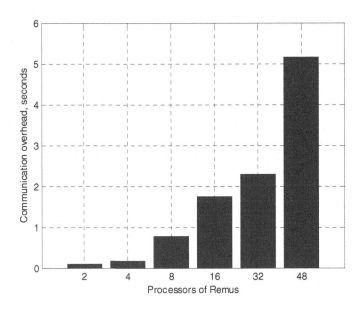

Fig. 4. Average (on 5 scenarios) communication overhead of the parallel algorithm

5 Conclusions

The development of the parallel batch pattern back propagation training algorithm of multilayer perceptron with two hidden layers of neurons and the research of its parallelization efficiency for the large-scale data classification task on many-core high performance computing system are presented in this paper. The model of multilayer perceptron and the batch pattern back propagation training algorithm are theoretically described. The algorithmic description of the parallel batch pattern training method is presented. Our results show high parallelization efficiency of the developed algorithm on many-core system with 48 CPUs: the parallelization efficiency is (i) 96-98% for the three bigger parallelization scenarios with 240, 320 and 400 neurons in the first hidden layer and 150, 200 and 250 neurons in the second hidden layer and 30000, 40000 and 50000 training patterns respectively and (ii) 72-78% for the two smaller parallelization scenarios with 80 and 160 neurons in the first hidden layer and 50 and 100 neurons in the second hidden layer and 10000 and 20000 training patterns respectively.

The future direction of our research can be considered as investigation of parallel training algorithms for other multilayer architectures, in particular convolution neural network.

Acknowledgment. The authors acknowledge the support of administration and technical staff of the Innovative Computing Lab at the University of Tennessee, USA for the use of their high performance computing infrastructure.

References

1. Haykin, S.: Neural Networks and Learning Machines, 3rd edn. Prentice Hall, New Jersey (2008)
2. De Llano, R.M., Bosque, J.L.: Study of Neural Net Training Methods in Parallel and Distributed Architectures. Future Generation Computer Systems 26(2), 183–190 (2010)
3. Čerňanský, M.: Training Recurrent Neural Network Using Multistream Extended Kalman Filter on Multicore Processor and Cuda Enabled Graphic Processor Unit. In: Alippi, C., Polycarpou, M., Panayiotou, C., Ellinas, G. (eds.) ICANN 2009, Part I. LNCS, vol. 5768, pp. 381–390. Springer, Heidelberg (2009)
4. Lotrič, U., Dobnikar, A.: Parallel Implementations of Recurrent Neural Network Learning. In: Kolehmainen, M., Toivanen, P., Beliczynski, B. (eds.) ICANNGA 2009. LNCS, vol. 5495, pp. 99–108. Springer, Heidelberg (2009)
5. Parallel Grid-aware Library for Neural Network Training, http://uweb.deis.unical.it/turchenko/research-projects/pagalinnet/
6. Turchenko, V., Grandinetti, L.: Scalability of Enhanced Parallel Batch Pattern BP Training Algorithm on General-Purpose Supercomputers. In: de Leon F. de Carvalho, A.P., Rodríguez-González, S., De Paz Santana, J.F., Corchado Rodríguez, J.M. (eds.) Distributed Computing and Artificial Intelligence. AISC, vol. 79, pp. 525–532. Springer, Heidelberg (2010)

7. Turchenko, V., Grandinetti, L.: Parallel Batch Pattern BP Training Algorithm of Recurrent Neural Network. In: 14th IEEE International Conference on Intelligent Engineering Systems, Las Palmas of Gran Canaria, Spain, pp. 25–30 (2010)

8. Turchenko, V., Bosilca, G., Bouteiller, A., Dongarra, J.: Efficient Parallelization of Batch Pattern Training Algorithm on Many-core and Cluster Architectures. In: 7th IEEE International Conference on Intelligent Data Acquisition and Advanced Computing Systems, Berlin, Germany, pp. 692–698 (2013)

9. Turchenko, V., Golovko, V., Sachenko, A.: Parallel Training Algorithm for Radial Basis Function Neural Network. In: 7th International Conference on Neural Networks and Artificial Intelligence, Minsk, Belarus, pp. 47–51 (2012)

10. Funahashi, K.: On the Approximate Realization of Continuous Mappings by Neural Network. Neural Networks 2, 183–192 (1989)

11. Hornik, K., Stinchcombe, M., White, H.: Multilayer Feedforward Networks are Universal Approximators. Neural Networks 2, 359–366 (1989)

12. The MNIST Database of Handwritten Digits, http://yann.lecun.com/exdb/mnist/

13. Hinton, G.E., Osindero, S., Teh, Y.: A Fast Learning Algorithm for Deep Belief Nets. Neural Computation 18, 1527–1554 (2006)

14. Golovko, V., Galushkin, A.: Neural Networks: Training, Models and Applications. Radiotechnika, Moscow (2001) (in Russian)

15. Turchenko, V., Grandinetti, L., Bosilca, G., Dongarra, J.: Improvement of Parallelization Efficiency of Batch Pattern BP Training Algorithm Using Open MPI. Procedia Computer Science 1(1), 525–533 (2010)

16. Open MPI: Open Source High Performance Computing, http://www.open-mpi.org/

17. Turchenko, V.: Scalability of Parallel Batch Pattern Neural Network Training Algorithm. Artificial Intelligence. Journal of Institute of Artificial Intelligence of National Academy of Sciences of Ukraine 2, 144–150 (2009)

Author Index